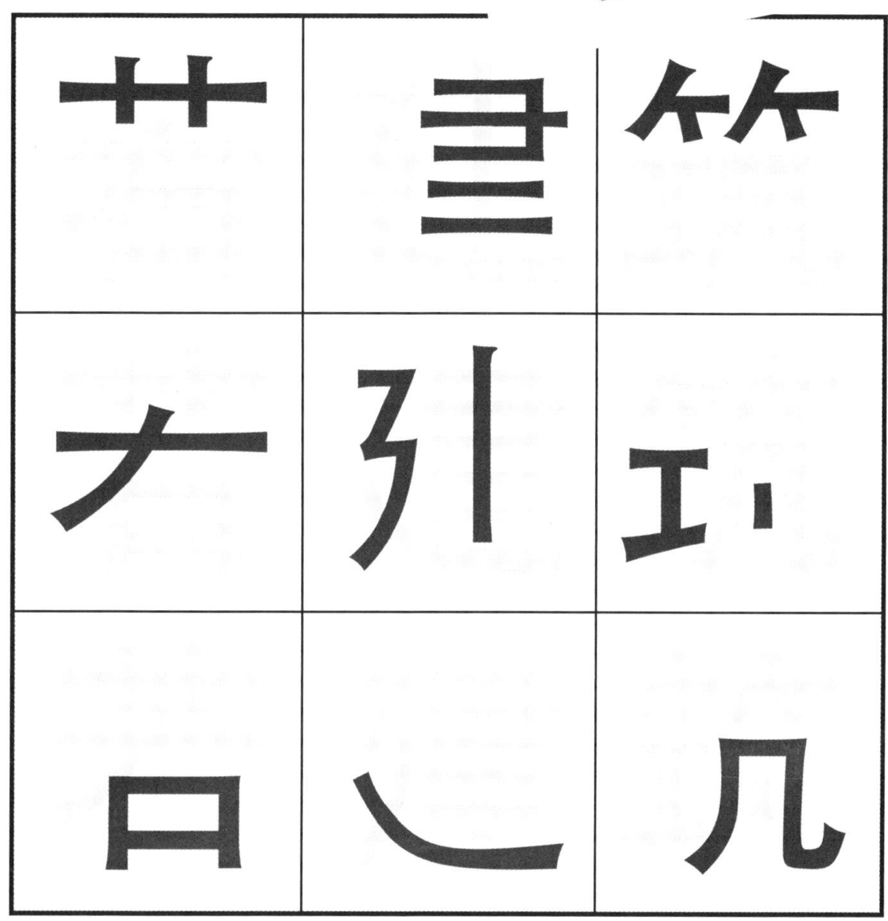

若建筑 PAN-ARCHITECTURE

王绍森 著

CHINA CITY PRESS

图书在版编目（CIP）数据

若建筑 = PAN-ARCHITECTURE / 王绍森著 . —北京：中国城市出版社，2021.9
ISBN 978-7-5074-3392-0

Ⅰ.①若… Ⅱ.①王… Ⅲ.①建筑设计 Ⅳ.① TU2

中国版本图书馆 CIP 数据核字（2021）第 185725 号

本书作者以多年实践经验提出若建筑的概念：简单地说，就是与建筑相关的图式记录和图示表达。建筑师通过对周边环境的观察记录抽象，或事件图示记录，形成抽象或具象的图示记录，作为与建筑有关的城乡环境的记录，成为训练建筑师观察分析的基础，通常成为图示的速写记录。图示思维作为建筑师的创作和训练表达方式。可以是一种图示思维也可是设计综合，这种成果就形成了既是建筑又非建筑的图示成果，就称为若建筑。

责任编辑：杨　琪　陈　桦
责任校对：张　颖

若 建 筑
PAN-ARCHITECTURE
王绍森　著
*
中国城市出版社出版、发行（北京海淀三里河路9号）
各地新华书店、建筑书店经销
北京雅盈中佳图文设计公司制版
北京中科印刷有限公司印刷
*
开本：787 毫米 ×1092 毫米　1/16　印张：$14^1/_2$　字数：265 千字
2021 年 11 月第一版　2021 年 11 月第一次印刷
定价：**79.00** 元
ISBN 978-7-5074-3392-0
（904377）

厦门大学“双一流”建设子项目“建筑文化传统与传承”成果
国家自然科学基金（51878581）基于复杂系统论的现代闽台地域建筑设计方法提升研究

前　言

一、何为若建筑 Pan-Architecture Concept

何为若建筑？简单地说，就是与建筑相关的图式记录和图示表达。

What is Pan-Architecture? Briefly, Pan-Architecture is recording and expressing architecture with images.

作为建筑师，我们应该有“图—途—图”三个阶段专业语言的训练和成果。

For architects, there could be 3 professional training and expression stages: Image-Diagram-Vision.

第一个字，“图”就是建筑师通过对周边环境的观察记录抽象，或事件图式记录，形成抽象或具象的图式记录，作为与建筑有关的城乡环境的记录，成为训练建筑师观察分析的基础，通常成为图式的速写记录。

Image that records surrounding environment and events, is one of the basic observe and analysis training methods for architects, usually results as sketches.

第二个字，“途”即途径，图示思维作为建筑师的创作和训练表达方式。通常情况下，实际的建筑设计受到各种各样的限制，很难以展开，但是在我们个人的图示思维中，可以抛弃多种的制约，然后根据自己平时建筑师的积累和分析的基础上形成综合表现，它可以是一种图示思维，也可以是设计综合，这种成果就形成了既是建筑又非建筑的图示成果，所以我们把它称为若建筑。

Diagram, a method in architectural design and training. Actually, architectural design is limited because of various restrictions, and is difficult to start. Despite of these restrictions, architects could start from imagination, experiences and analysis, draw images of comprehensive thoughts and design, in architectural or non–architectural ways, and that could be called Pan–Architecture diagram.

第三个字，"图"即目的，我们通过这一类的图式记录和图示表达的训练，能够把一个建筑师训练成有敏锐的观察力和抽象分析能力和设计综合能力，综合判断力，"若建筑"的成果最终可提升设计能力和激发设计启迪，其中的一部分也都可以转化为真实的建筑、城市、景观等具体设计。

Vision is the result. By this type of image recording and expression training, architects could increase their ability of observation, judgement, analysis, and their design which could also be inspired by Pan–Architecture. Part of the vision could eventually be transformed to actual architectural, urban and landscape design.

二、创作过程 Design Process

我们知道任何事物都是由复杂的系统构成，若建筑的形成也是由系统来构成的，首先触点始发，其次协调关系，然后是编织结构，最后形成系统。

As we know, everything consists of complex systems. Pan–architecture is also the same. The process includes 4 steps: Start point. Relationship coordination. Structure construction. System Generation.

触点始发：第一个点是有可能是一个场景，一个事物，也可能是建筑师自己的心境，然后引爆我们从一点开始进行创作，进行记录和抽象发展，这实际上是若建筑最重要的一个开头的方向。

Start point: The beginning of the first step may be a scene, an object or a thought in the architect's mind. These small events could encourage architects to record, develop and design. It is the most important starting point of Pan–Architecture.

协调关系：若建筑在发展的过程中间，我们会协调很多关系，比如形式与材质，路径与空间，人工与自然，然后场景与意境的关系等。如此之类平时积累的各种关系素材在此均得以再现和重组。

Relationship coordination: During the process of Pan–Architecture design, many relationships should be coordinated, such as form and material, path and space, artificial environment and nature, scene and artistic concept etc. These accumulated materials could be reviewed and reorganized in relationship coordination process.

编织结构：结构十分重要，所谓的结构就是在我们的思维过程中间所形成的牢固趋势，它构成的空间关系和思维引导。同时通过这种结构的方式控制若建筑中的空间场景以及人能够体味愉悦的一些特别的引导。形成的结构在整个的发展过程中至关重要，控制总体趋势。

Structure construction：Structure is the strong trend in design process that combine all concepts together. It forms the spaces and expresses the thought. It controls the scene and ensure pleasant experience. Structure is very import in the whole process because it could control the trend.

形成系统：无论出发点是什么，在过程中研究了各种各样的关系，编织结构，最终的目的是为了形成整个系统，在这个系统里是整体意向，而这个系统本身是一个总体性的概括，是一个总体性的体现。

System generation: No matter what the start point is or how the relationships and structures are created, the purpose is to generate the entire system, and this system should be conceptual and integrated.

三、呈现特点 Features

相由心生：所谓相由心生，是指若建筑所呈现出来的图式，是有建筑师主观内心有形呈现，综合内心思维所流露出来的显型。一方面可以是日积月累的环境记忆，也可以是抽象思维的升华，又有内心心境的物化再现，这是综合重构的一种思维的图像。

Representation of thoughts: The schema of Pan–Architecture is the presentation of architect's thoughts. It could be the memories of the experiences accumulated in years, or the sublimation and materialization of thoughts. It is a kind of reconfigured image in architect's mind.

隐喻象征：在整个若建筑的图式中，图式有象征和隐喻的意义和模式，其中可以看到熟悉的和不熟悉的物件，在此可能与表达意义有所关联：比如拂尘关联风和自由、水池代表自然湖海，石块抽象伟岸山石，树木象征了森林，鼠标隐喻关系与控制……，而这些都是隐喻。当然，场景中更有人的介入，立、坐、行等代表人的各种行为的体现。

自然界所存在的事物，以及人所构成的东西，人和各要素形成一个综合的世界。

Metaphor and symbolization: The schema of Pan-Architecture has metaphor meanings and symbolization modes. Familiar or unfamiliar objects could be connected to special meanings. For instance, a horsetail whisk represents wind and freedom, a pond indicates lakes and oceans, a rock links to majestic mountains, a tree symbolized forests, and a mouse could be a metaphor for relationship & control. These are all examples of metaphor. Not only natural objects are in the scenes, human behavior, people standing, sitting or walking are also included in the schema. Natural objects and human behavior are combined to form an integrated world.

时空转换：若建筑中间，它不是完全意义上正常的绘画，它是思维上的自由和转换。自由体现为主观感性自由，理性逻辑和感性畅想的相结合。转换有两种方向，一个是时间的转换，可以共时性，即过去—现在—将来的多时共存；二是空间也可以转换，即思维角度的转换。空间场景是俯视和平视以及其他多角度的转换，多尺度之间的转换……这里特别得益于中国文化和中国园林的层次性体验。时空转换带来多样的场景，也会激发创作的可能性。

Transition of time and space: Pan-Architecture is a free transition of thoughts which is different with ordinary paintings. Free transition means combination of logic thinking & imagination. The transition happens in two ways: one leads to synchronicity, which means the coexistence of the past, the present and the future; the other is the transformation of space, which means that scenes in the drawings could be aerial view, human-eye view or any other view-points, in various scales. This concept benefits a lot from Chinese culture, especially Chinese Gardens. The transition of time and space would bring about diverse scenes and inspired works.

四、结果作用 IV Result

意达心境：若建筑目的不是为了画一张图，主要的目的有两种：第一是为了能够记录和抽象，记录表达我们周边的环境和内心的心境；第二，把理性思维和感性思维有机地结合起来，通过象征隐喻，时空转换的方式来完成一个创作的过程，把我们内心的世界以设计思维来表达。始于一点终于整体。通过触点引爆—协调关系—编织结构—形成系统的过程设计发展完成一个完整的意境表达，这里呈现的结果可以是：简单事物的抽象记录、中国园林的空间解构与抽象、城市—建筑—景观三位一体的设计、海滨小品的设计、设计思维的抽象表达等。如此之类的东西可以是设计，也可以是记录，丰富多彩。

Presentation of thoughts: Pan–Architecture is not only a drawing, it has two main purposes. First, it is an abstract record of the outer and inner world; the other is that it combines rational and perceptual thoughts, and completes the process of creation by symbolization, metaphor and transition of time–space. It may start from a single thought, but end as a whole system. Starting from an inspiration, it may go through a process of coordinating relationships and weaving structures among elements, and finally result as a completed system. The result could be abstract record of simple things, deconstruction and abstraction of Chinese Garden space, a design of urban, architectural and landscape scale, or a design of seaside feature, or abstract expression of thoughts. Such results could be either design or record, no specific form.

抽象激发：若建筑的作用大概分两个。第一个是训练设计抽象能力，因为我们周边的环境，纷繁多样，无形有形，建筑师需要抽象概括能力才能更好理解环境进行提升和设计，若建筑能够通过图示语言训练这种能力；第二个能力通过由点到面而形成一个系统的训练，形成意达心境的结果，中间过程和结果呈现中可以训练我们的思维，同样局部或整体设计结果也会成为我们在建筑创作中有机赋能激发。

Abstract inspiration: There could be two effects of Pan–Architecture. One is for abstract thinking training. Because of the diversity of the environment, architects need to summarize and improve it in an abstract way. Pan–Architecture offers training for this ability. The other is for training step by step, transforming ideas from mind to drawings.

Both the intermediate and the final results could be beneficial practice, and both the partial and the whole of the drawings could be the inspiration for architecture design.

厦门大学建筑与土木工程学院　院长　教授　建筑师

Dean，Professor，Architect

School of Architecture and Civil Engineering，Xiamen University

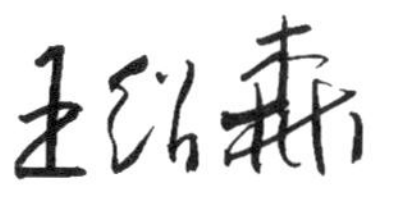

2020 年 8 月

目　录

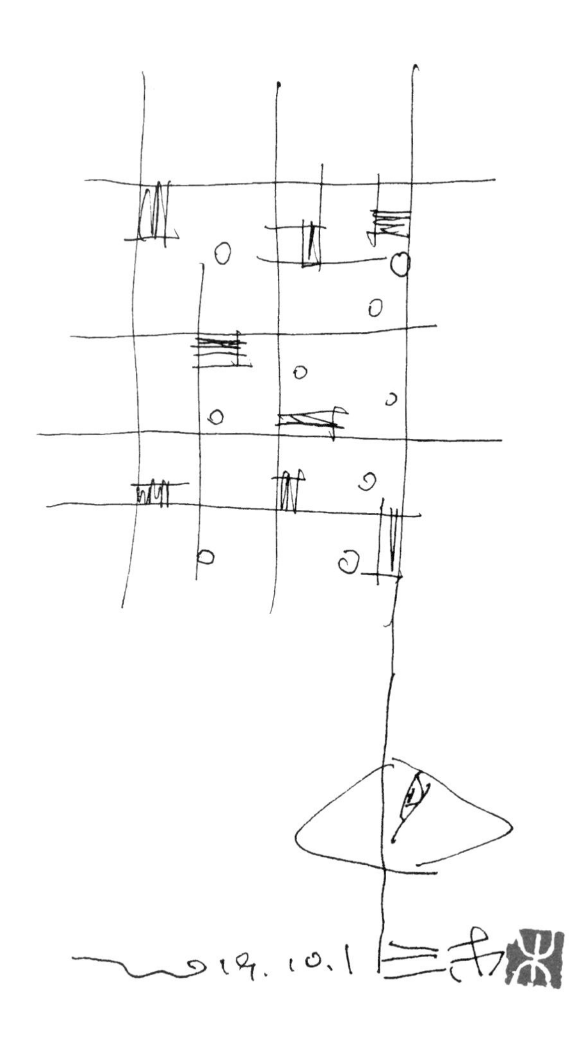
2014.10.1

1

若建筑——场所、情感、记忆

Pan-Architecture——place，emotion，memory

2008.5.10
Bali
www.thelayar.com

Bali

2006.6.27. OXFord
Sam. 三木

HOTEL
2006
LYON
SAM

2018.9.6

2019.11.25

2019.11.25. ROME
HoTel. VaLadier.

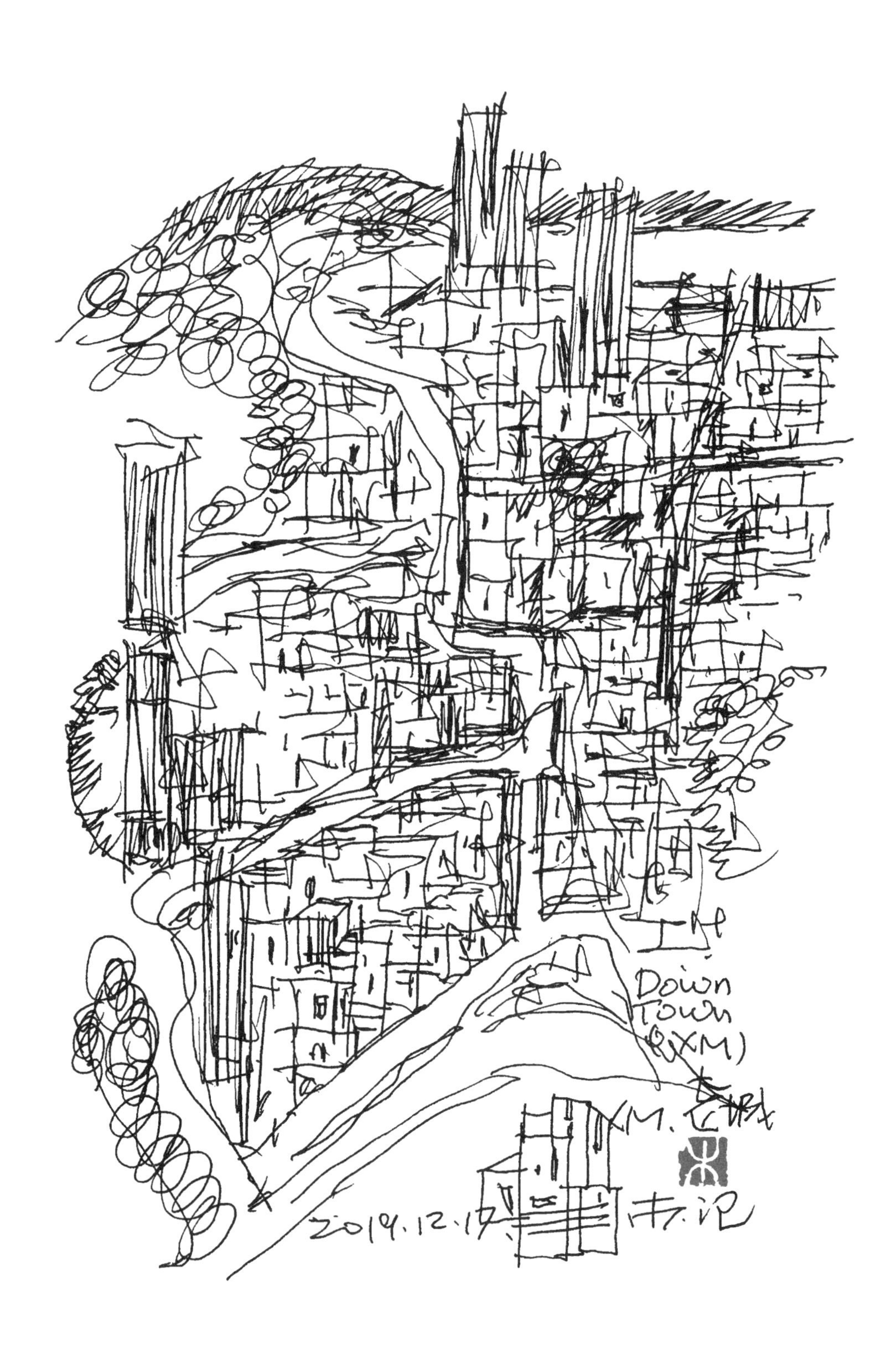
Down
Town
(XM)
2014.12.17

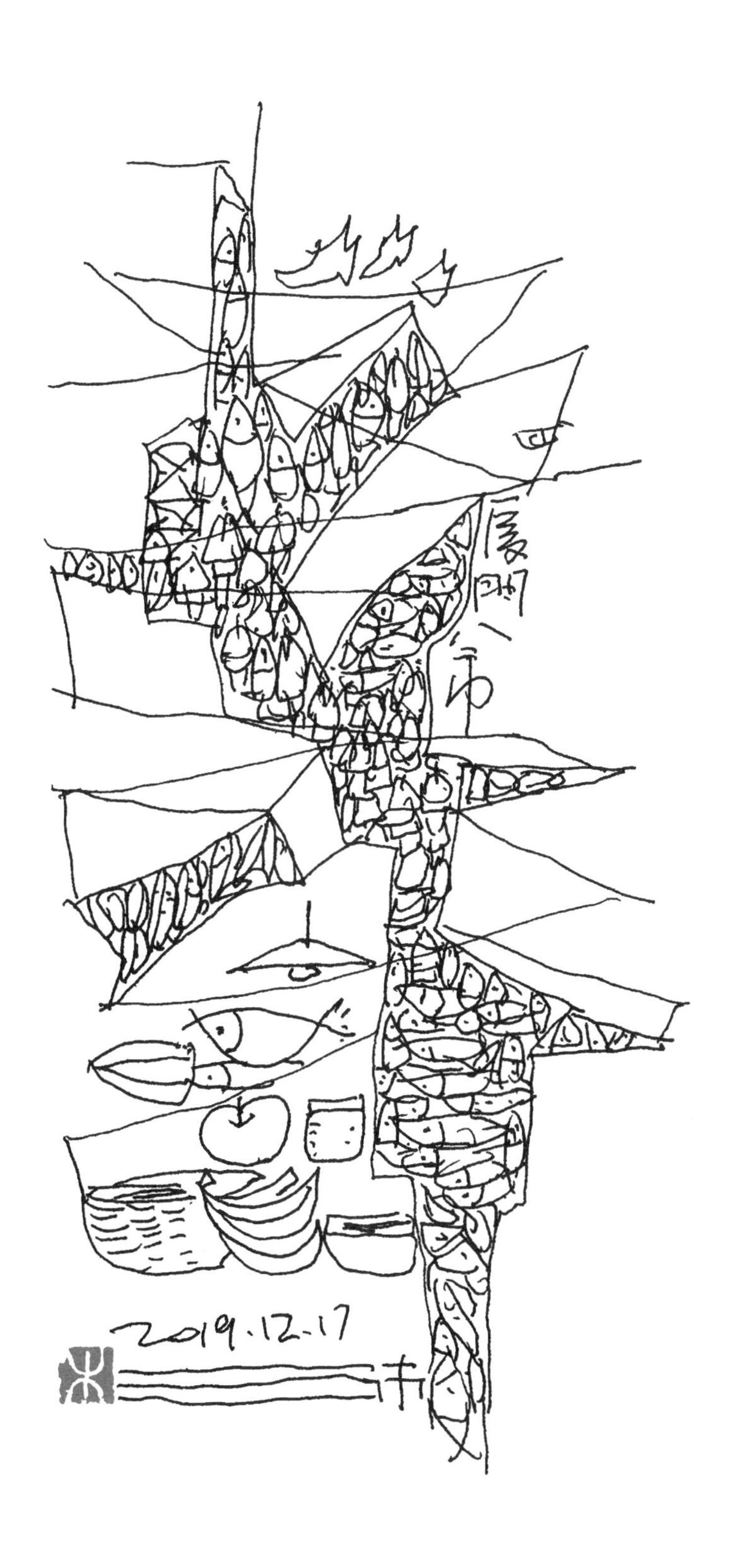
2019.12.17

2019.12.17

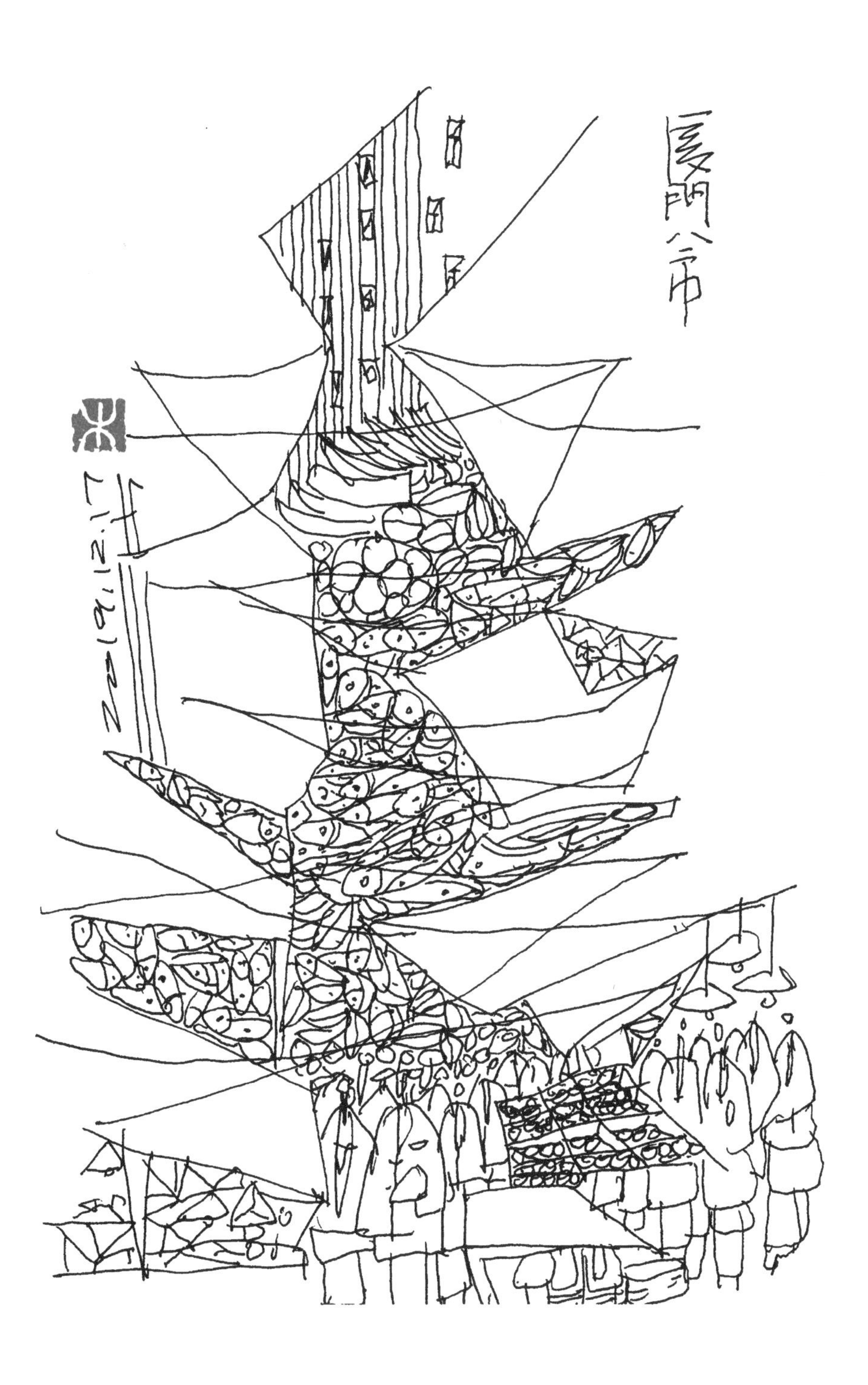

石頭厝.平潭.
2019.5.27

2019.5.27

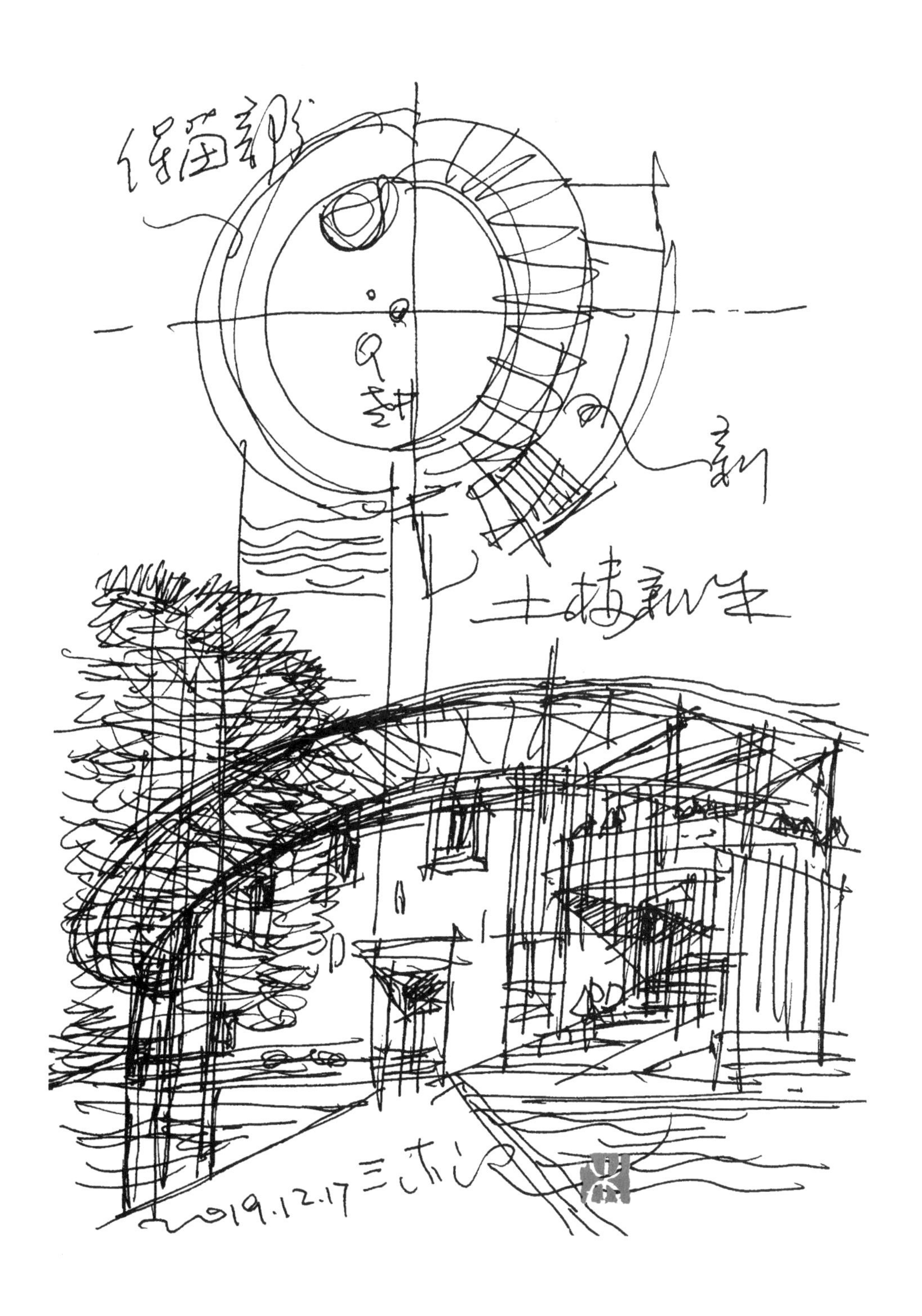
2019.12.17

2019.9.24

2

若建筑——是与建筑相关的图式记录和图示表达

Pan-Architecture is recording and expressing architecture with images.

2019.12.17
XMU

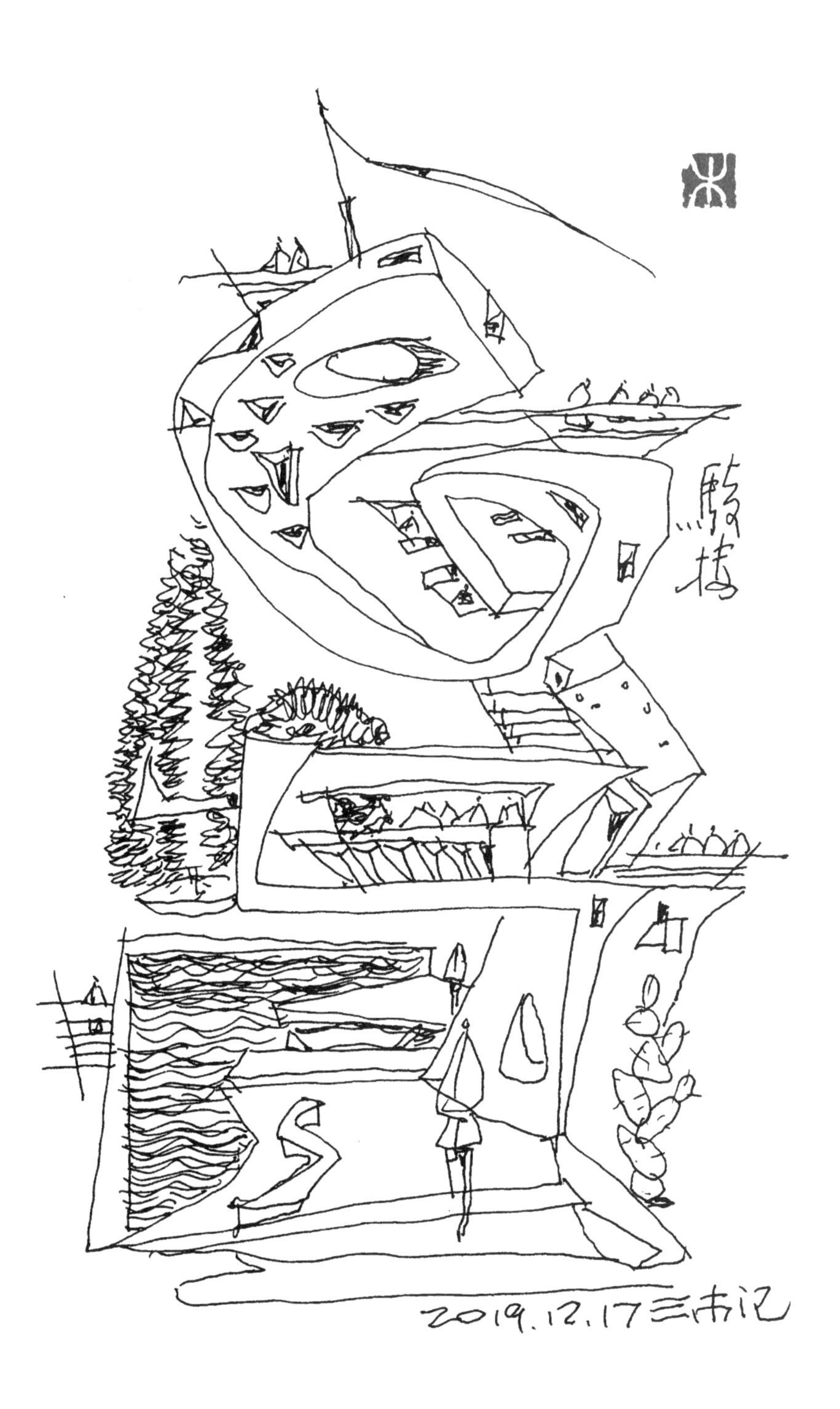
2019.12.17

2019.12.17

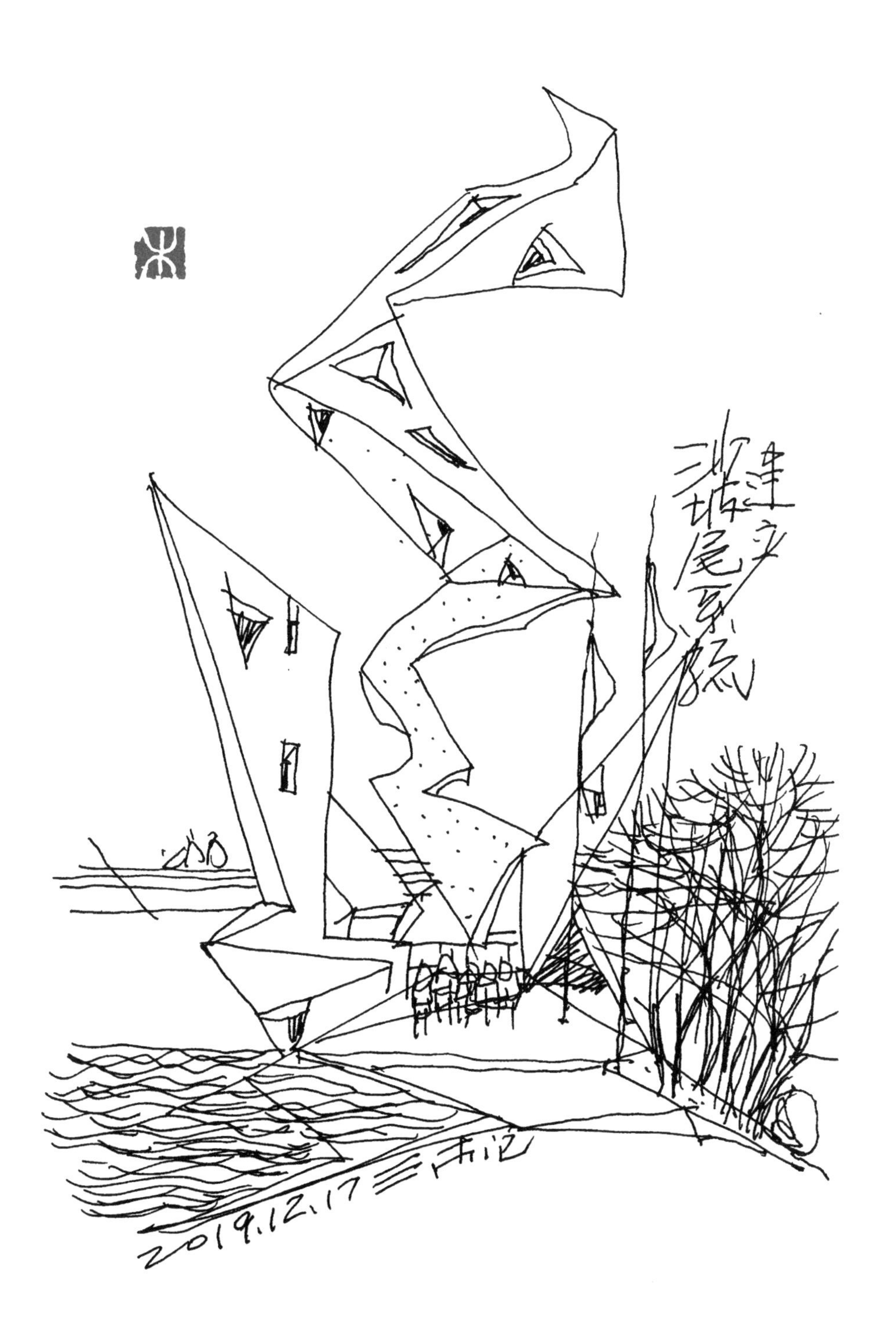
2019.12.17

风土
2020.1.1.

2020.1.6

2020.2.14

2020.4.24
XMU-UCA.

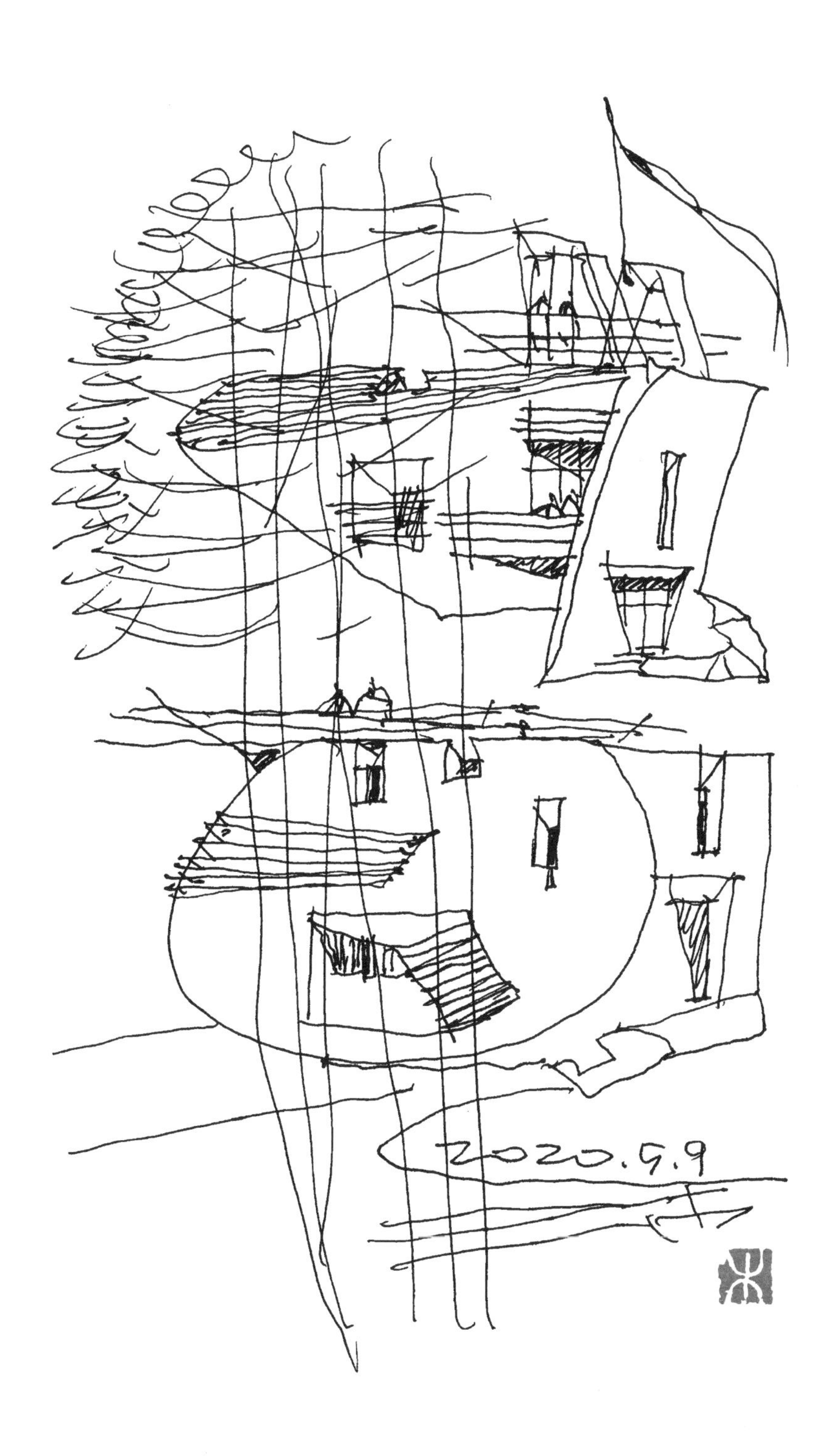
2020.5.9

2020.7.18

2020.17.23

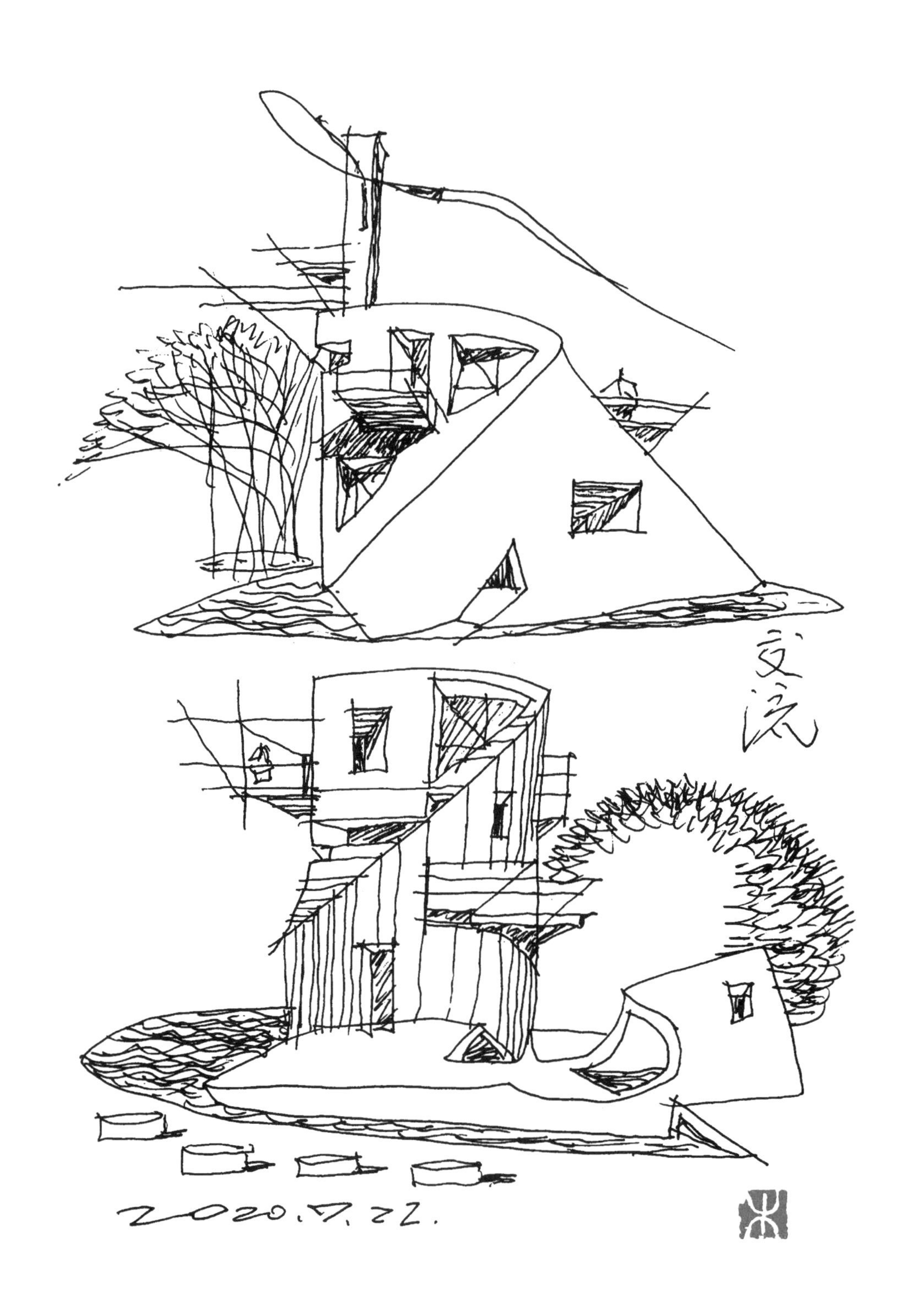
交流
2020.7.22.

2020.7.28

2020.7.28

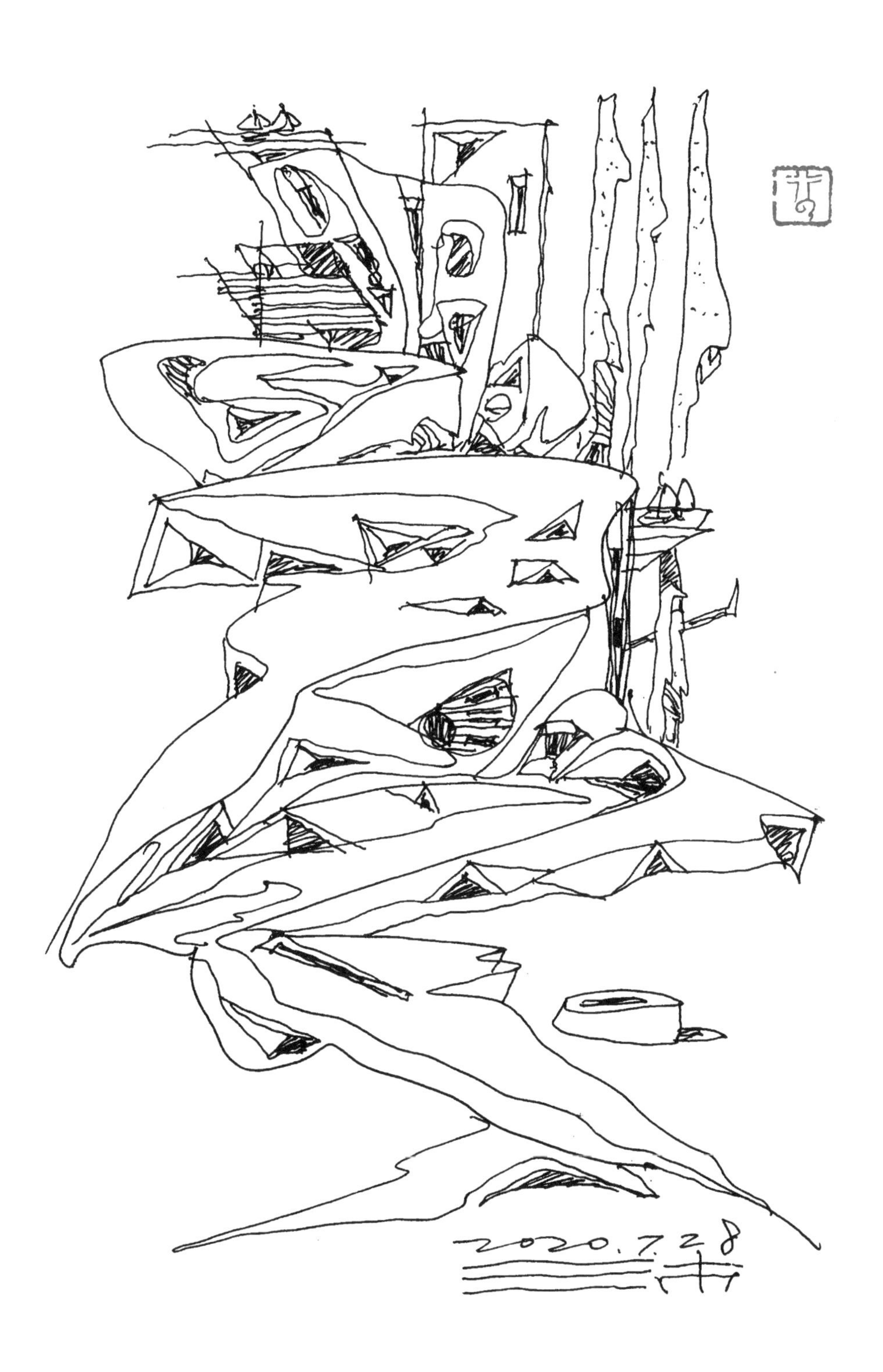
2020.7.28

2020.7.28,

2020.7.29

2020.7.8

2020.7.29

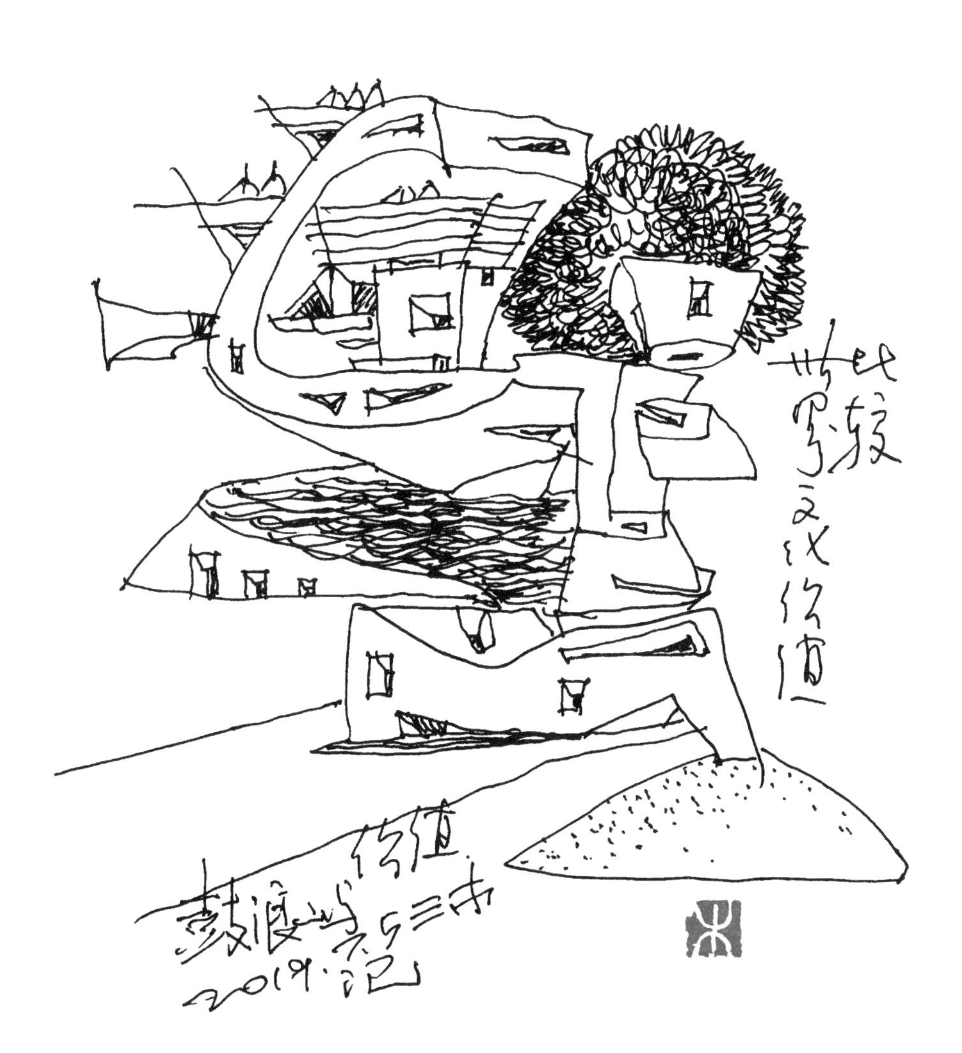

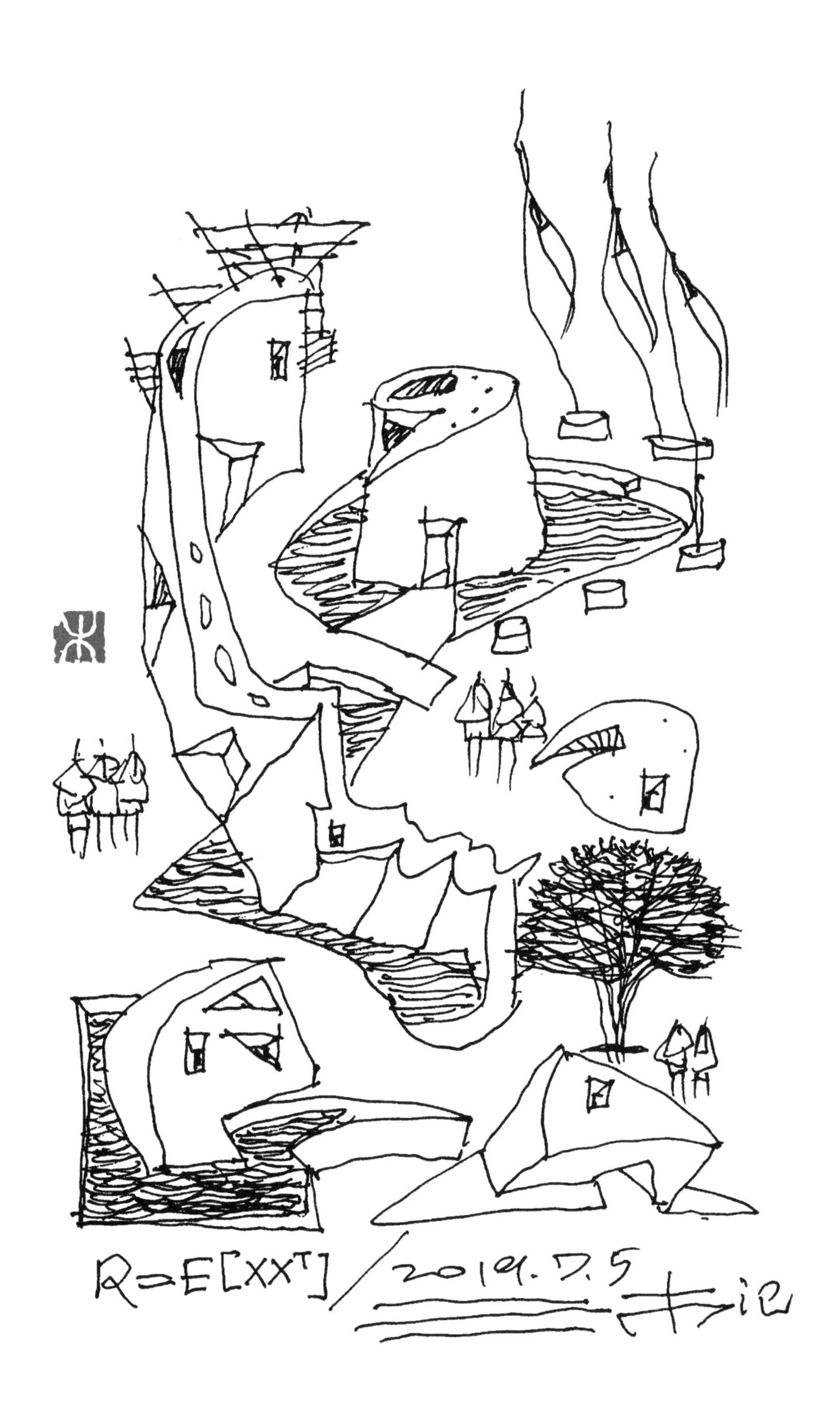
R=E[XXT]
2019.7.5

Peridynamic.
2019.7.5

中心广场
2019.5.27

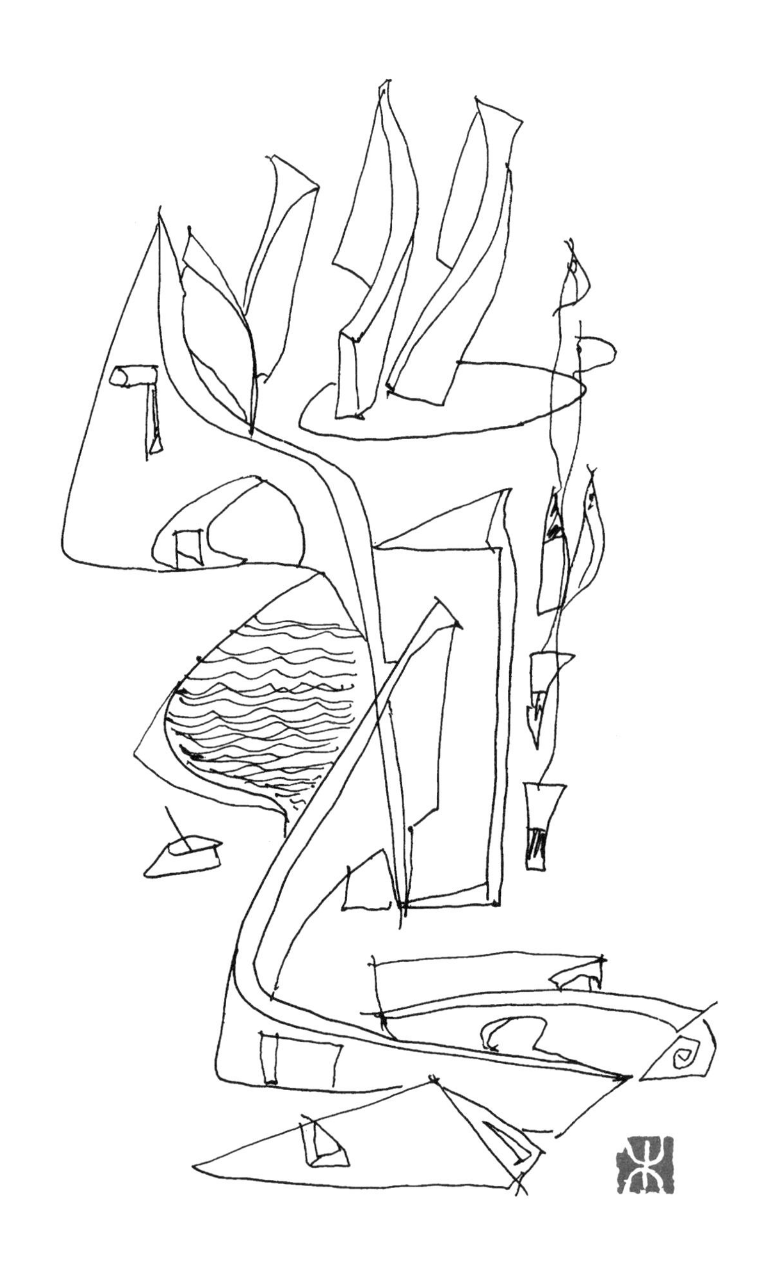

2019.9.24

2019.09.24

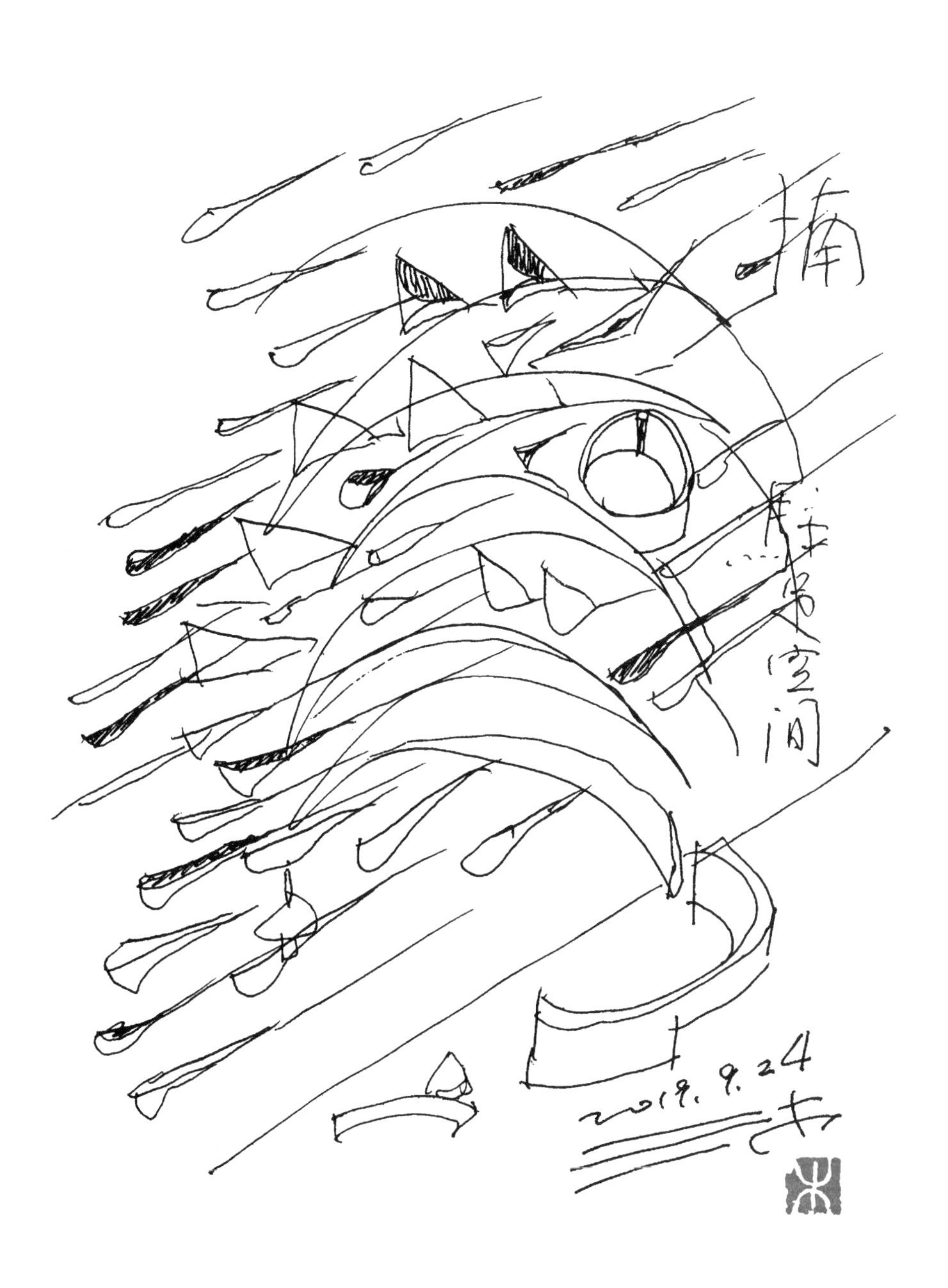
空间
2019.9.24

2019.6.9

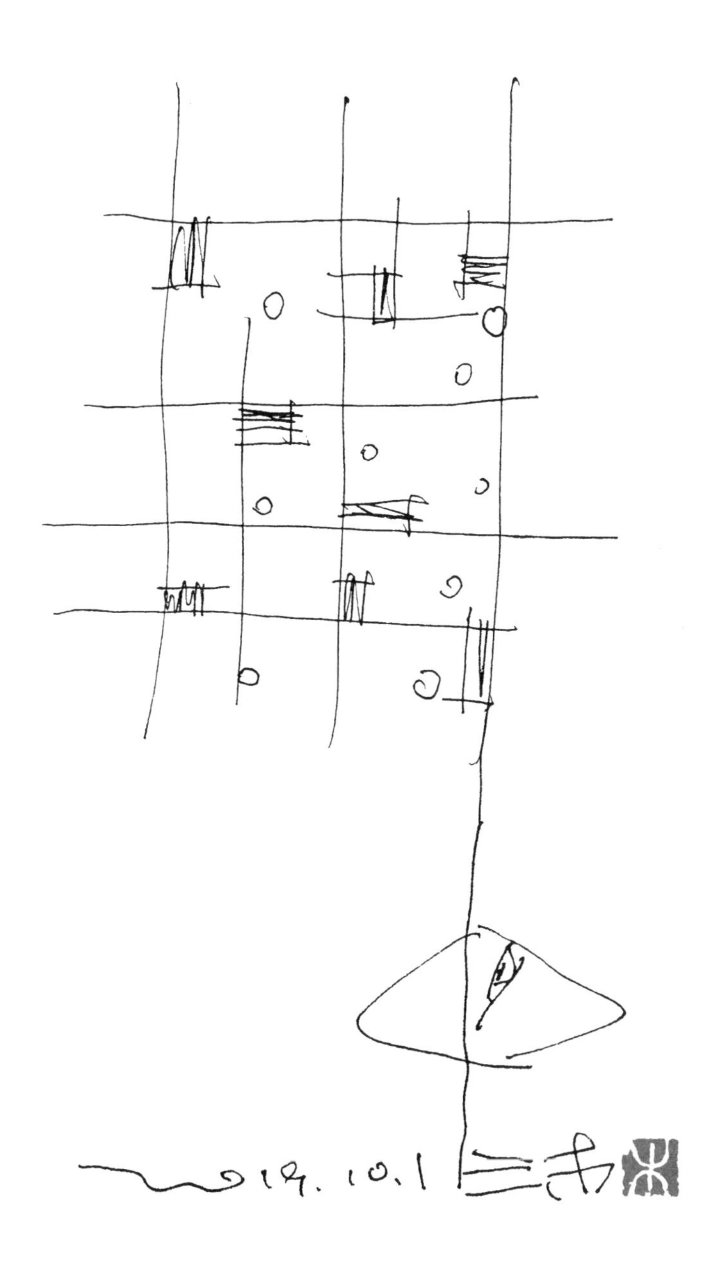
2019.10.1

2020.6.4.

2020.6.4

2020.6.4

2020. 6. 9.

2020.6.4

2020.6.4

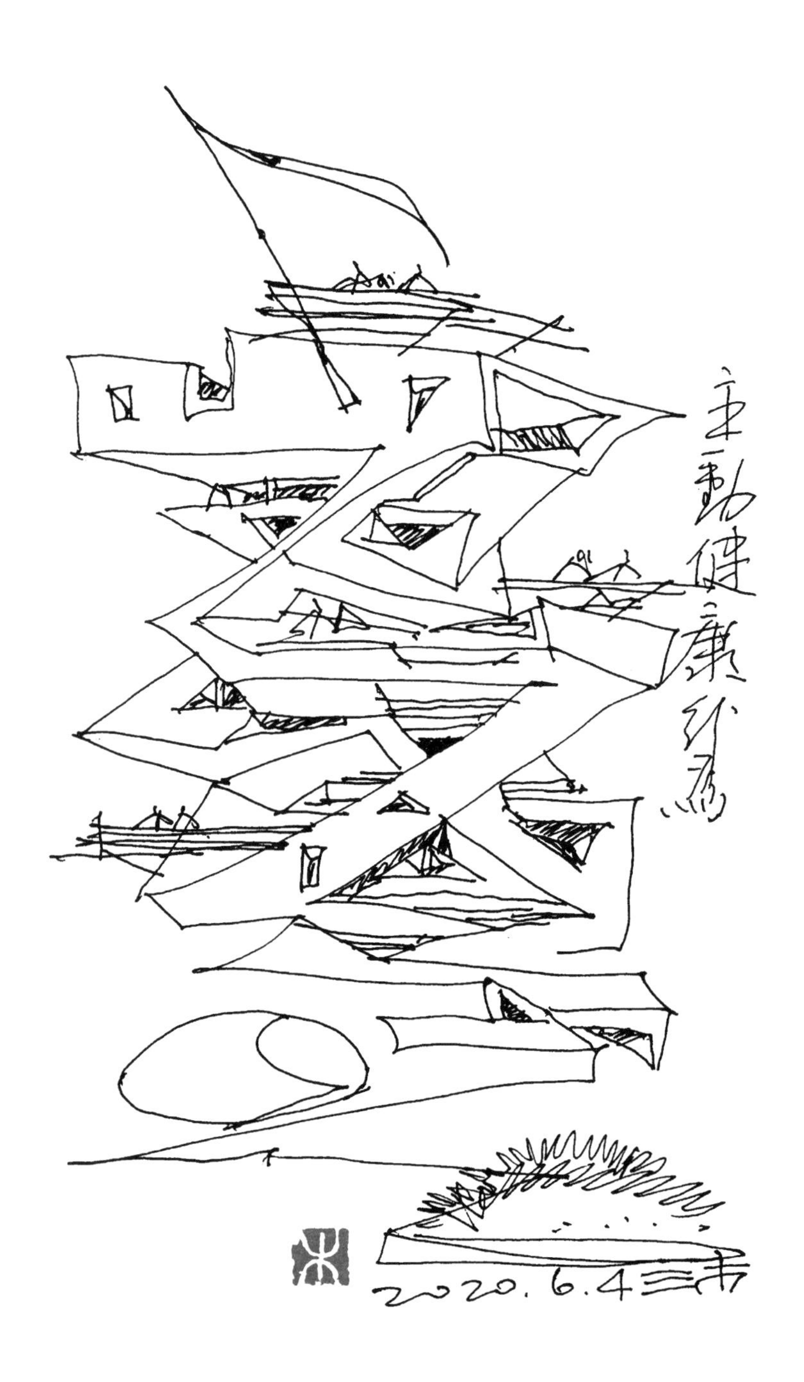
主動健康行為
2020.6.4

2020
6.4

TEA
2020 6.4

鄉村
2020 6.4

城市记忆
2020.
6.4.

2020.6.18

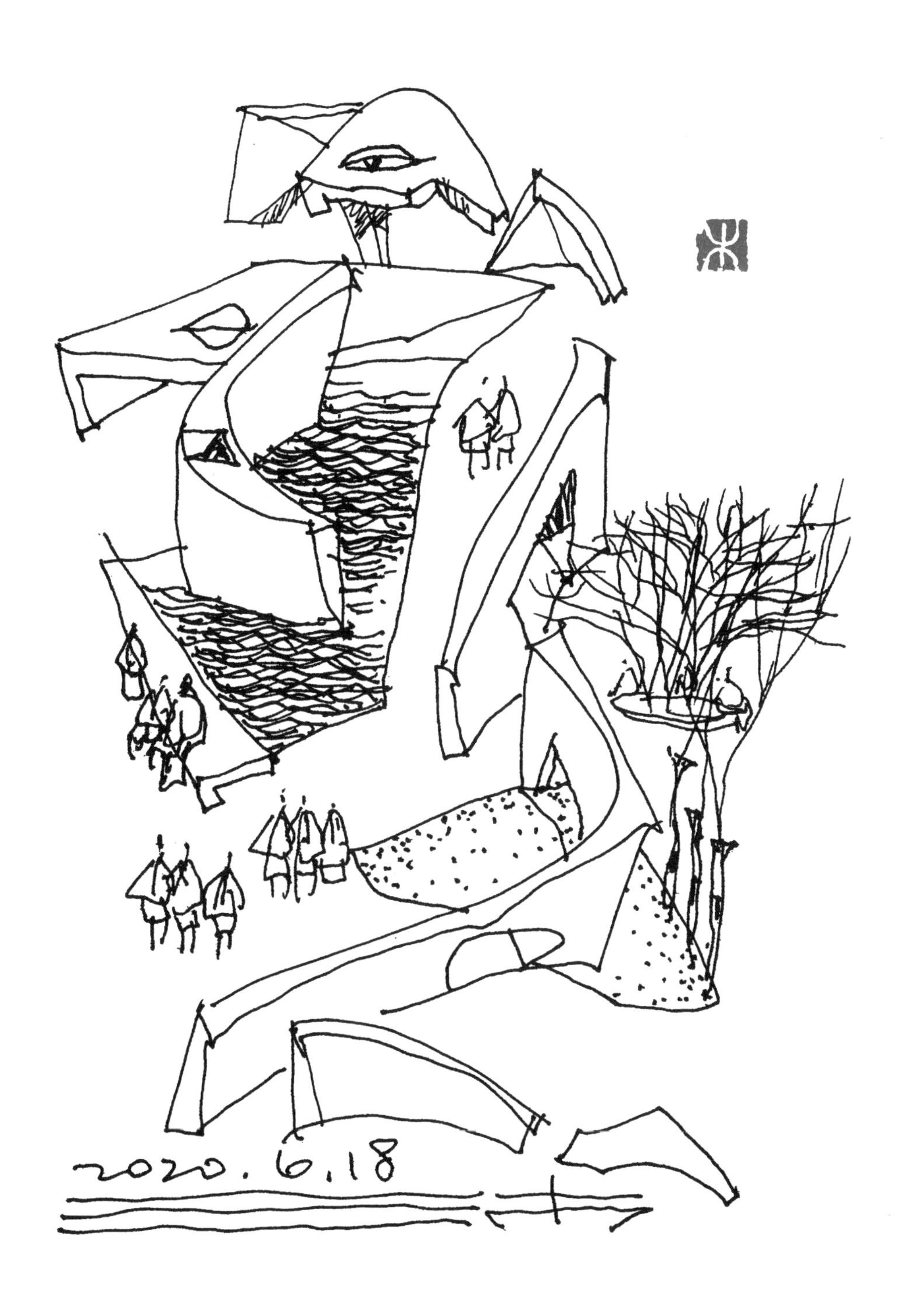
2020.6.18

2020.6.18

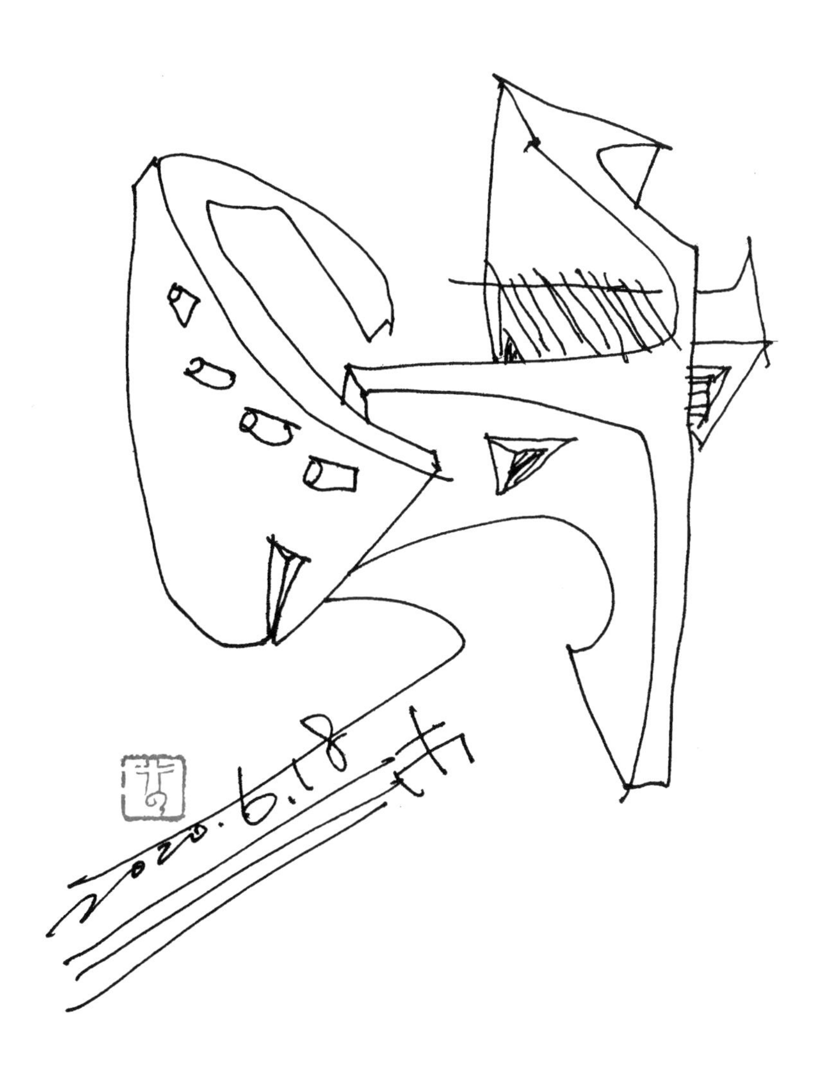
2020.6.18.

湖林
2020.6.18

2020.6.18

2020.(4+2).6.18.

Layers of City
2020.6.18

2020.6.28

2020.7.25.

表皮
2020.7.23

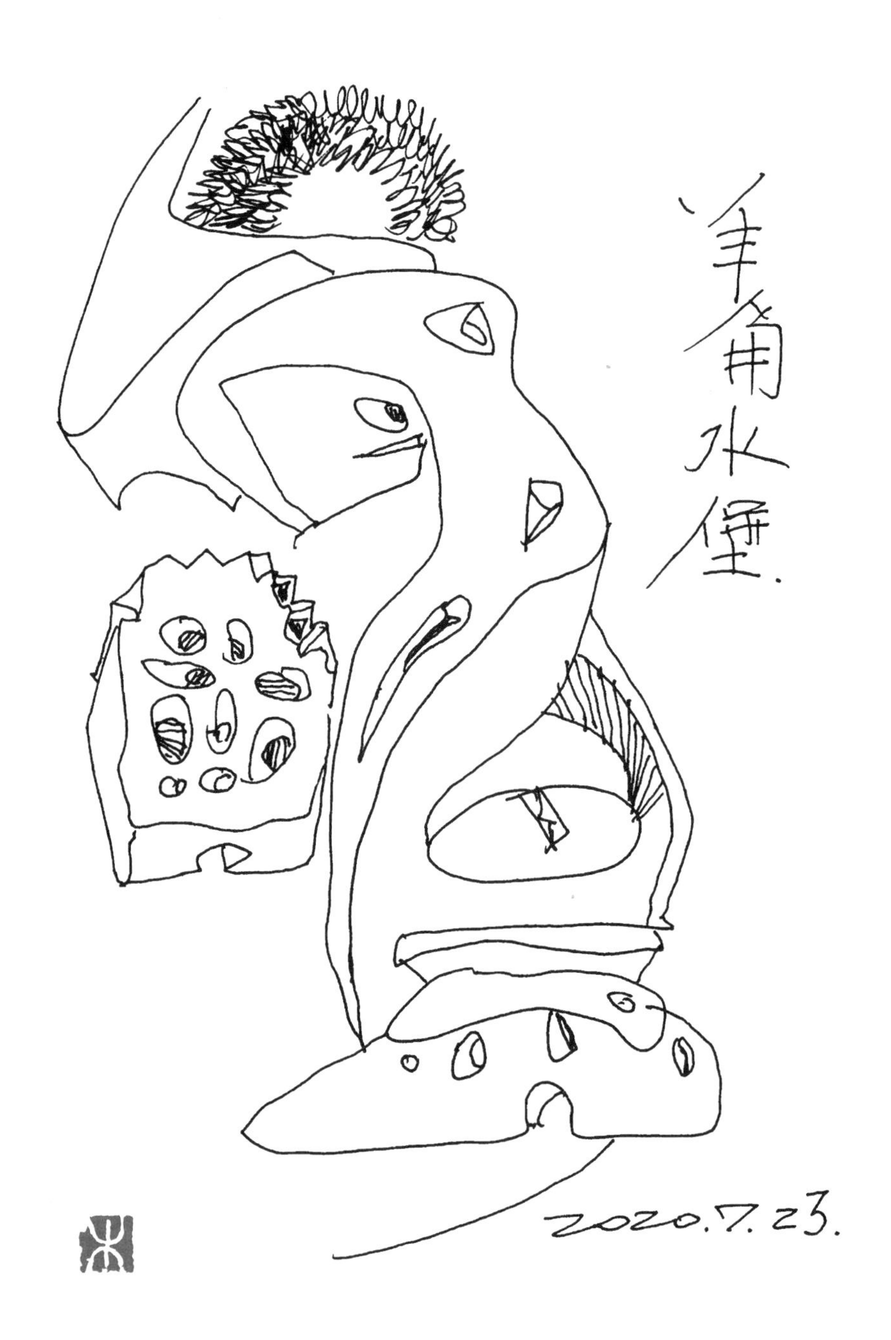
羊角水堡.
2020.7.23.

鼓浪屿.

微宅

2020.7.23

2020.7.23

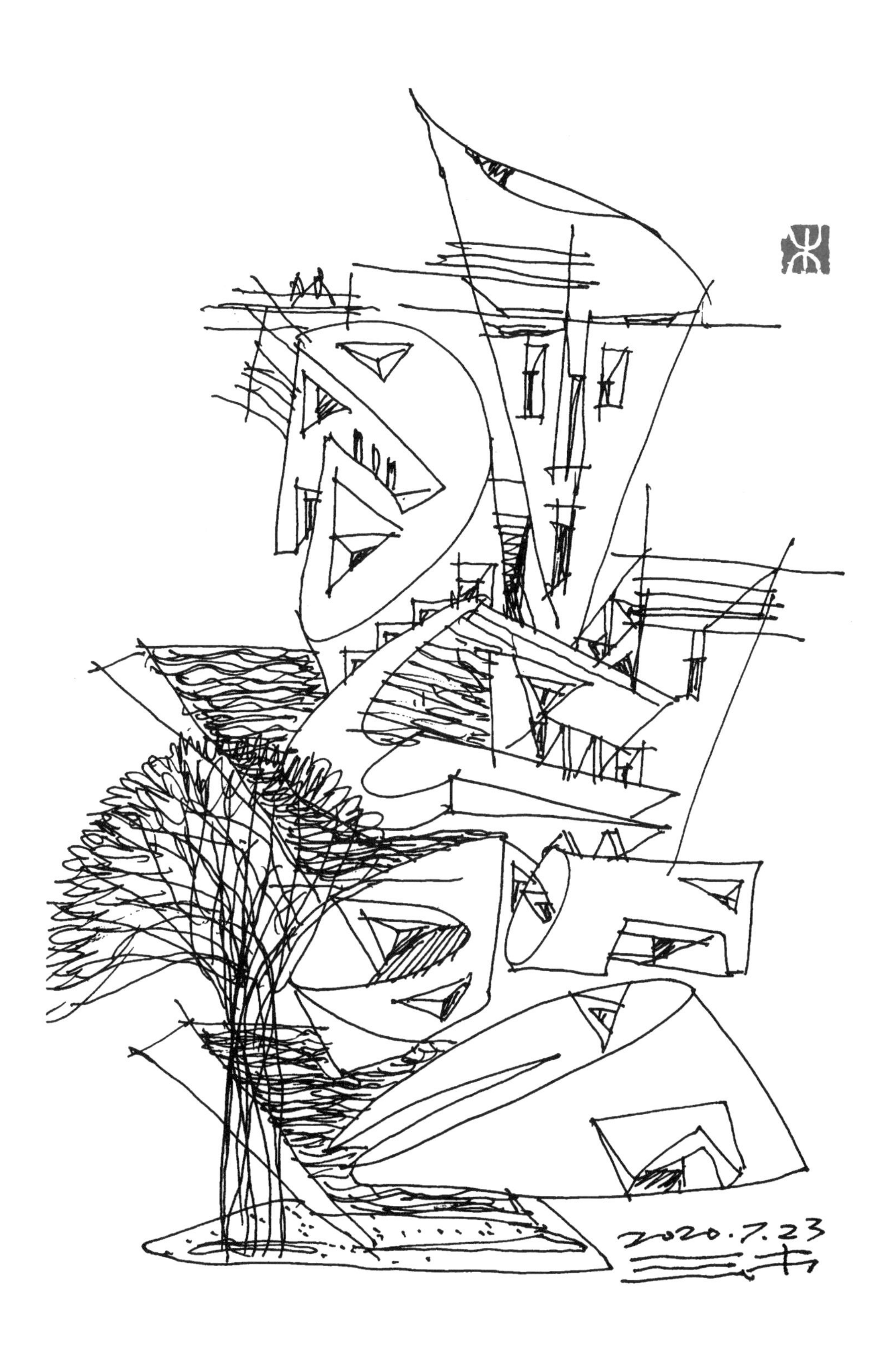
2020.7.23

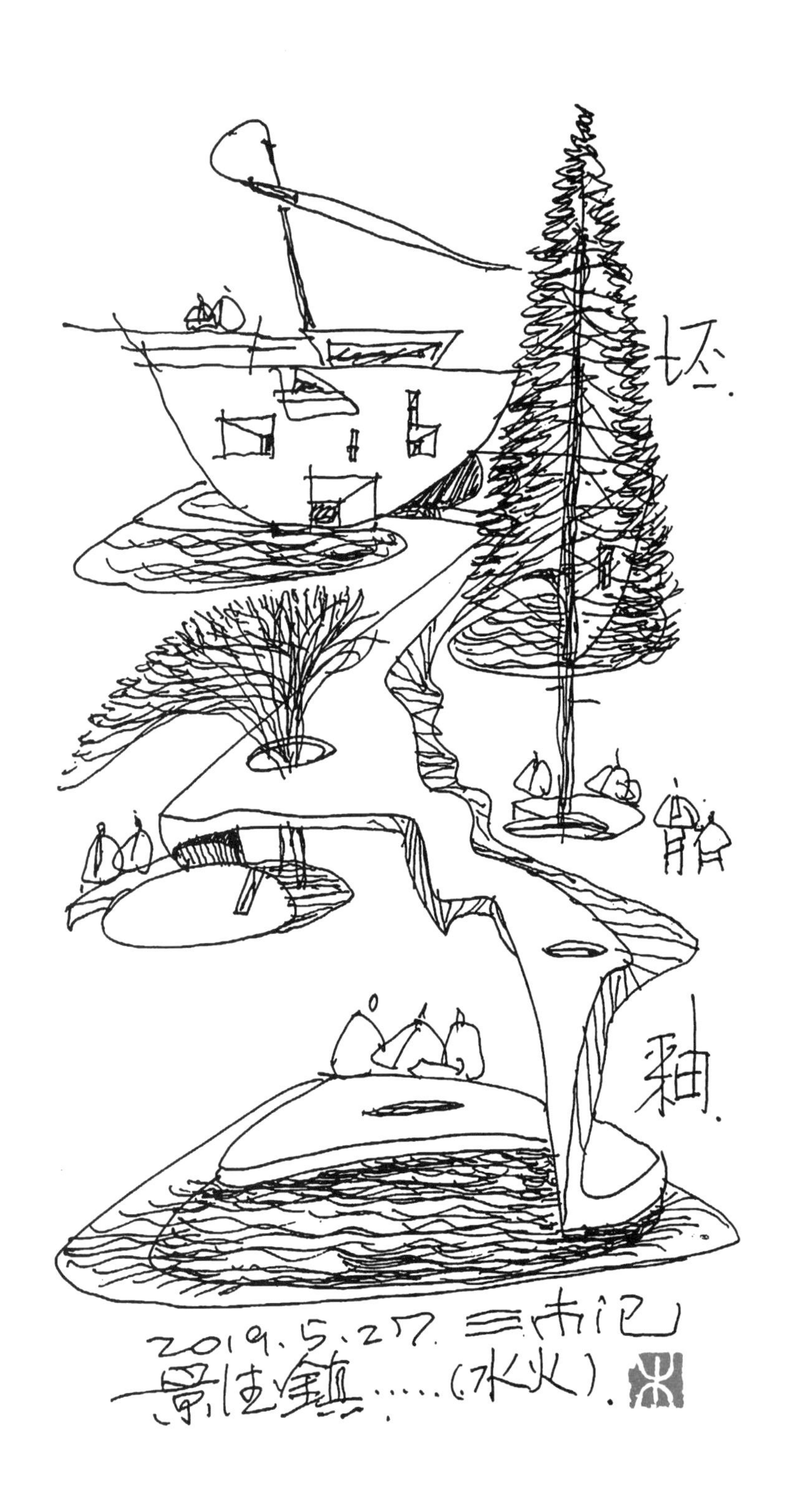
坯.
釉.
2019.5.27
景德镇……(水火).

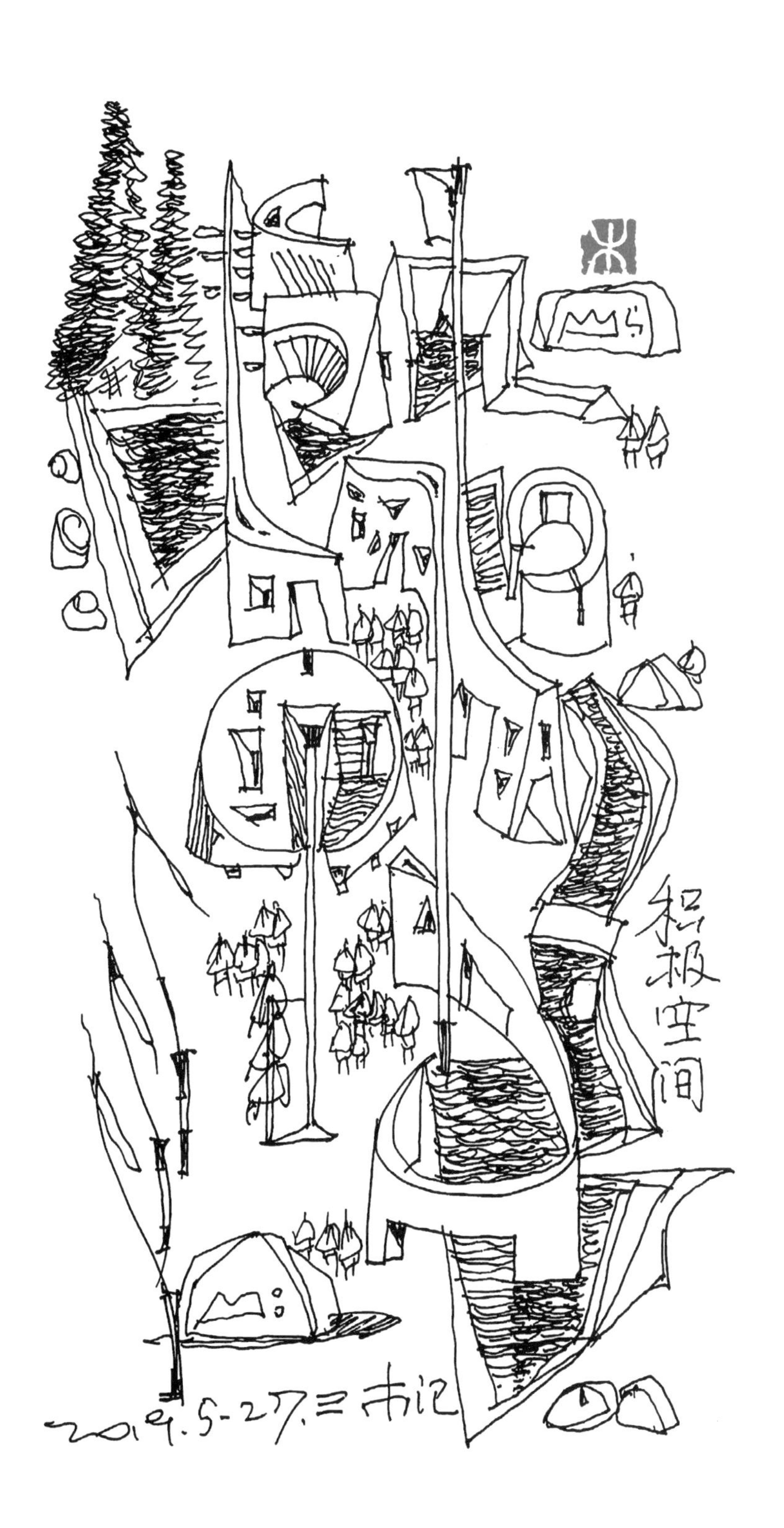
积极空间
2019.5-27

2019.5.27

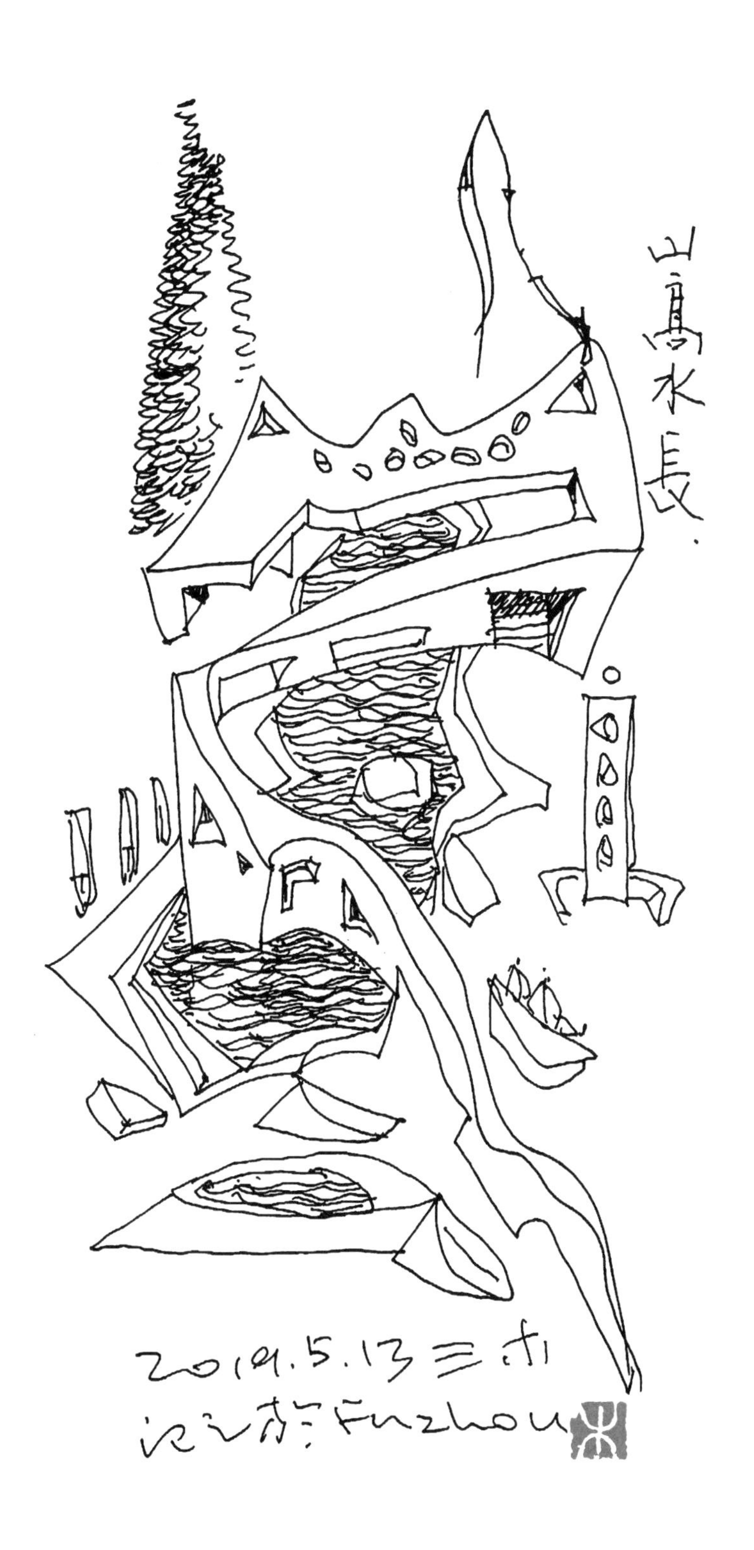
山高水長
2019.5.13
记于Fuzhou

海
码头
联系
商业
村
外地人
本地人
触媒
2019.5.27

2019.5.1

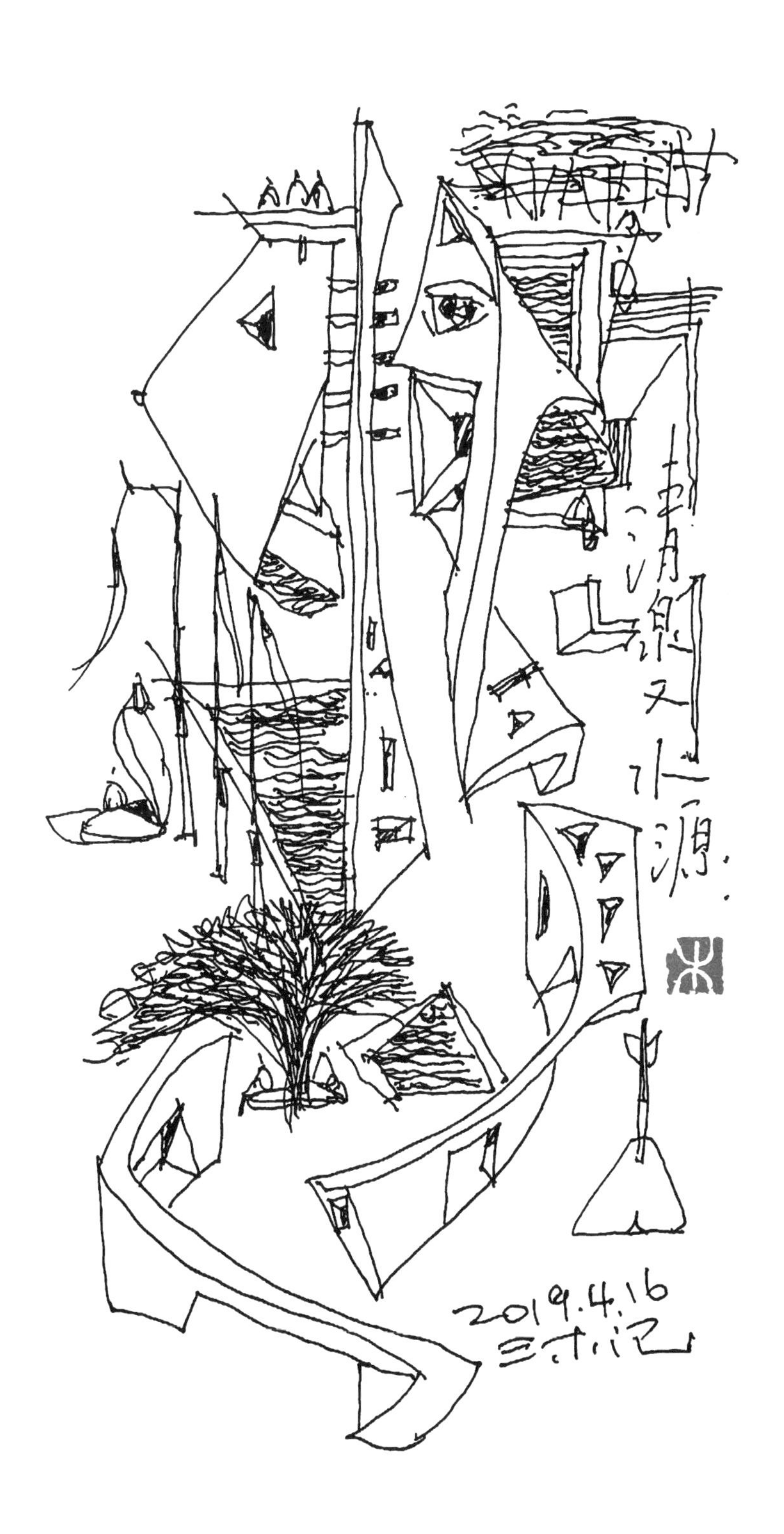
2019.4.16

2019.4.16

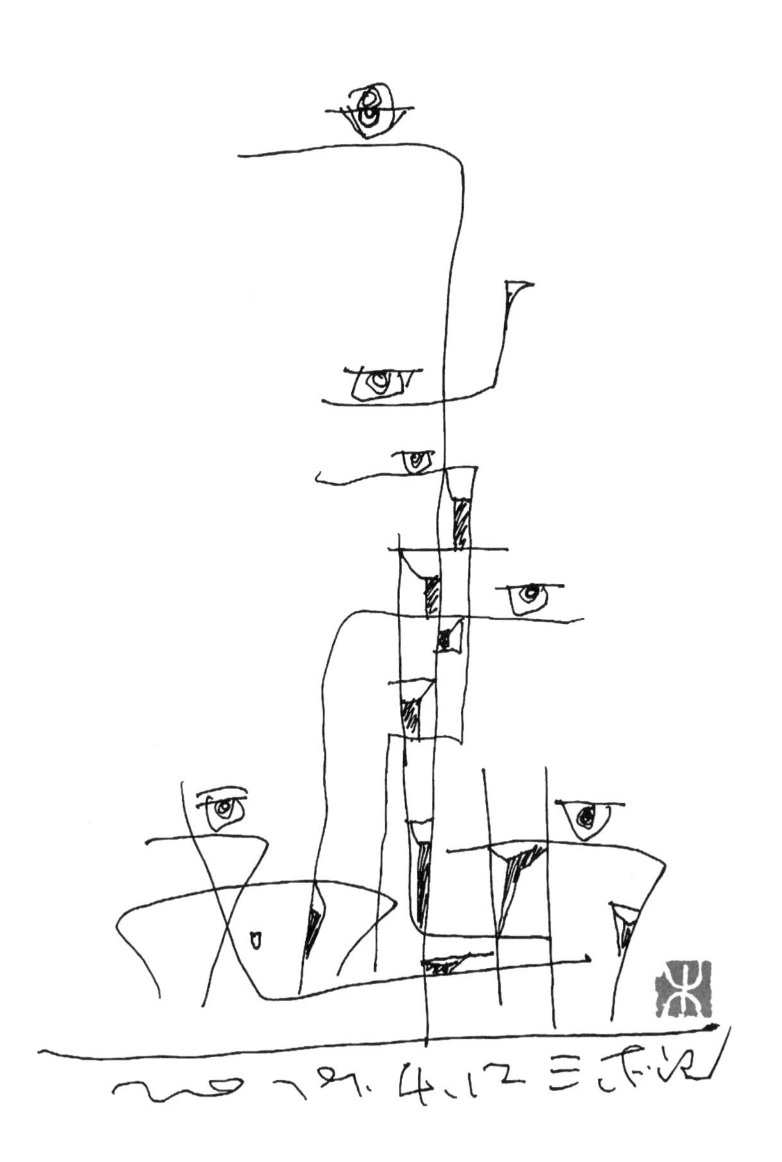

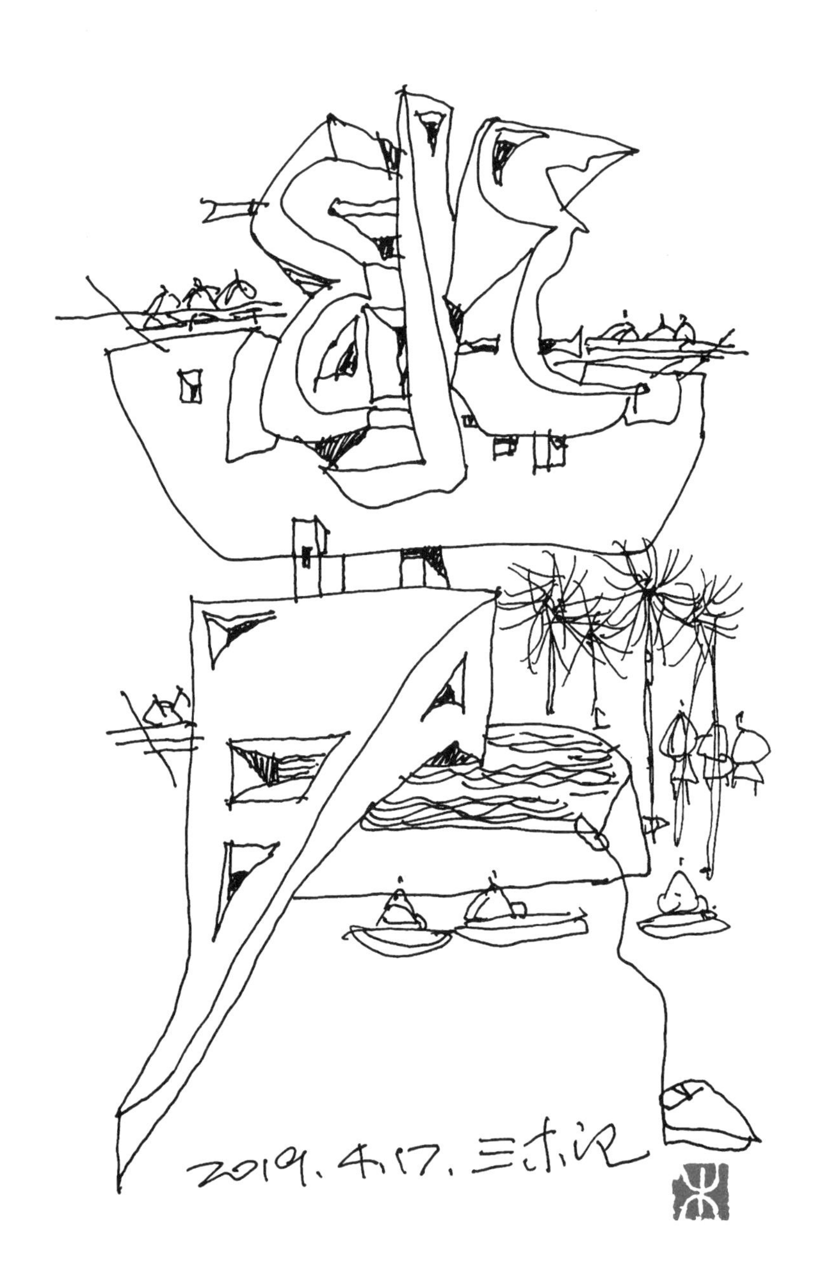
2019.4.17.

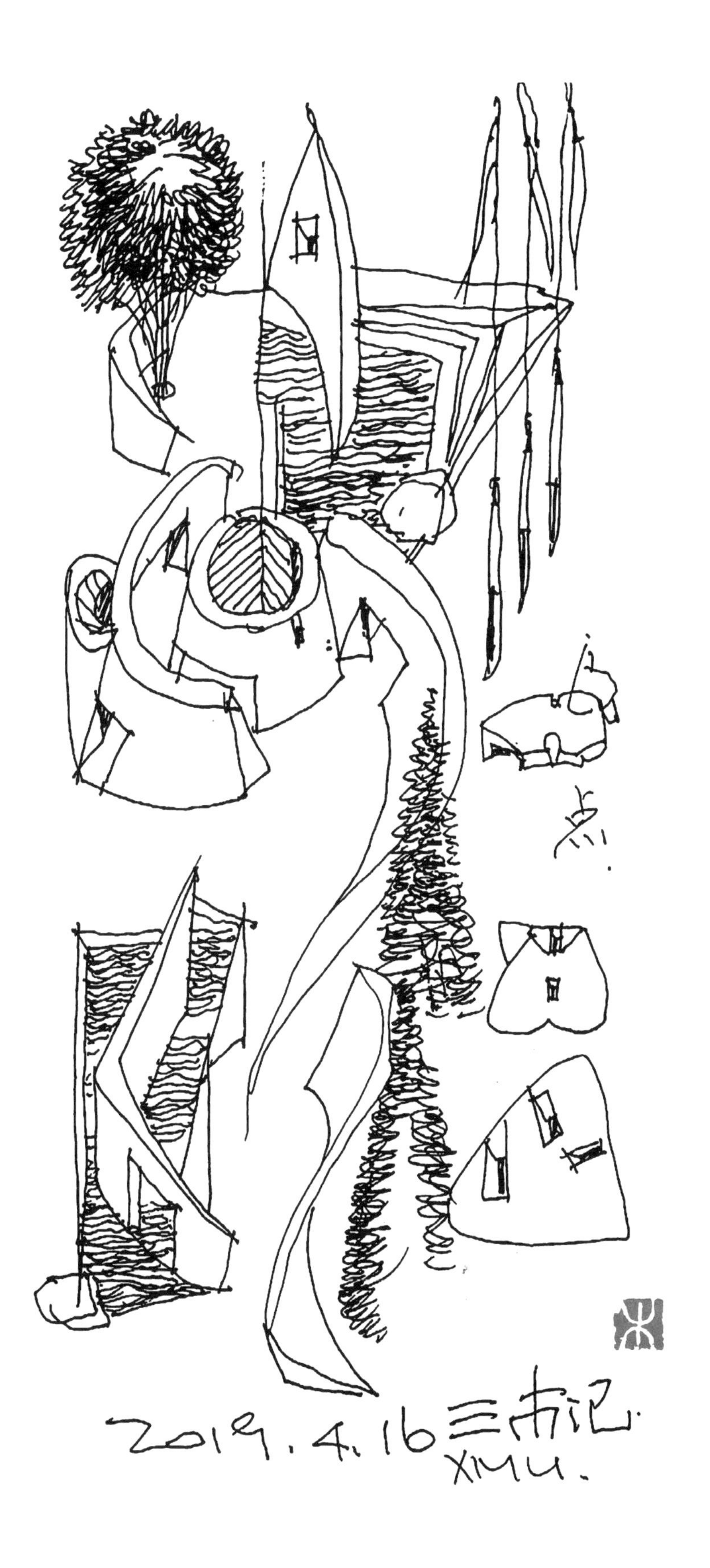
2019.4.16 三市记
XMU.

2019.4.16

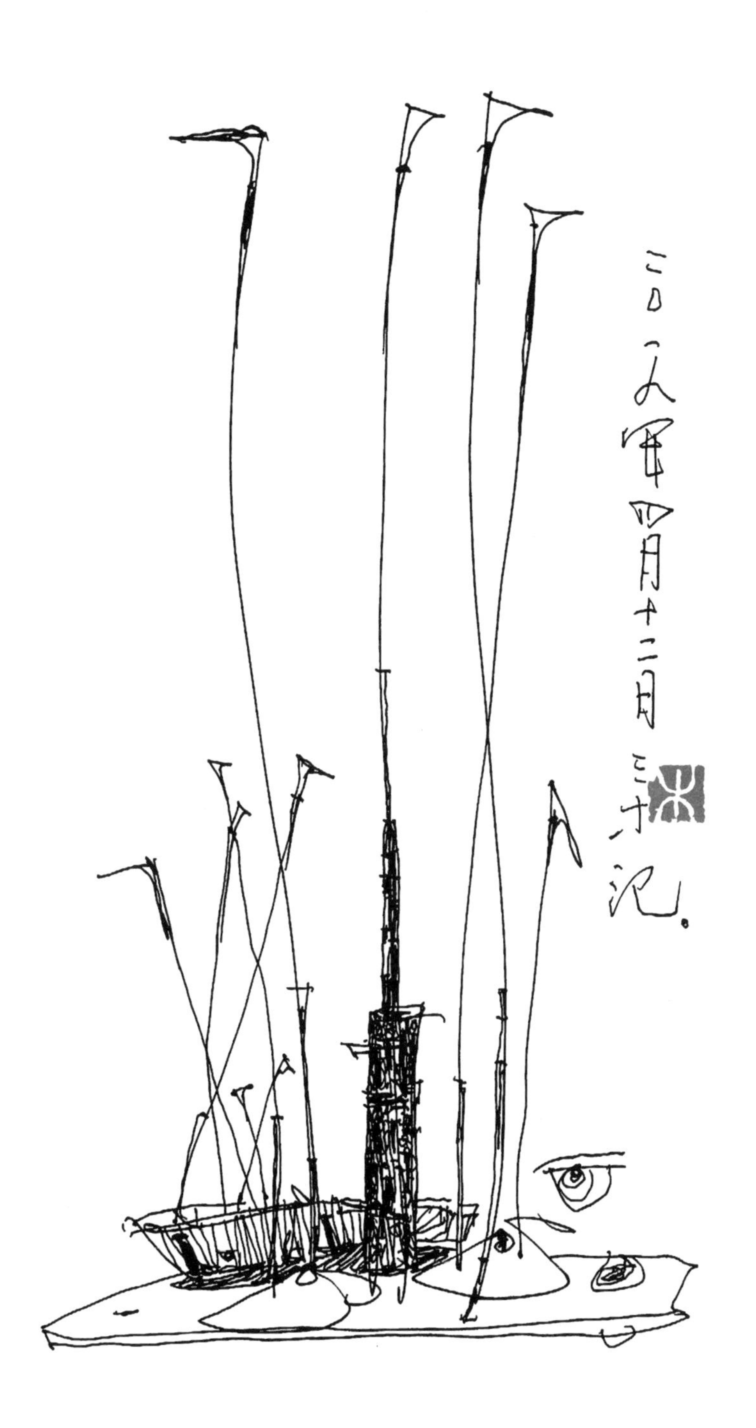
二〇一八年四月十二日

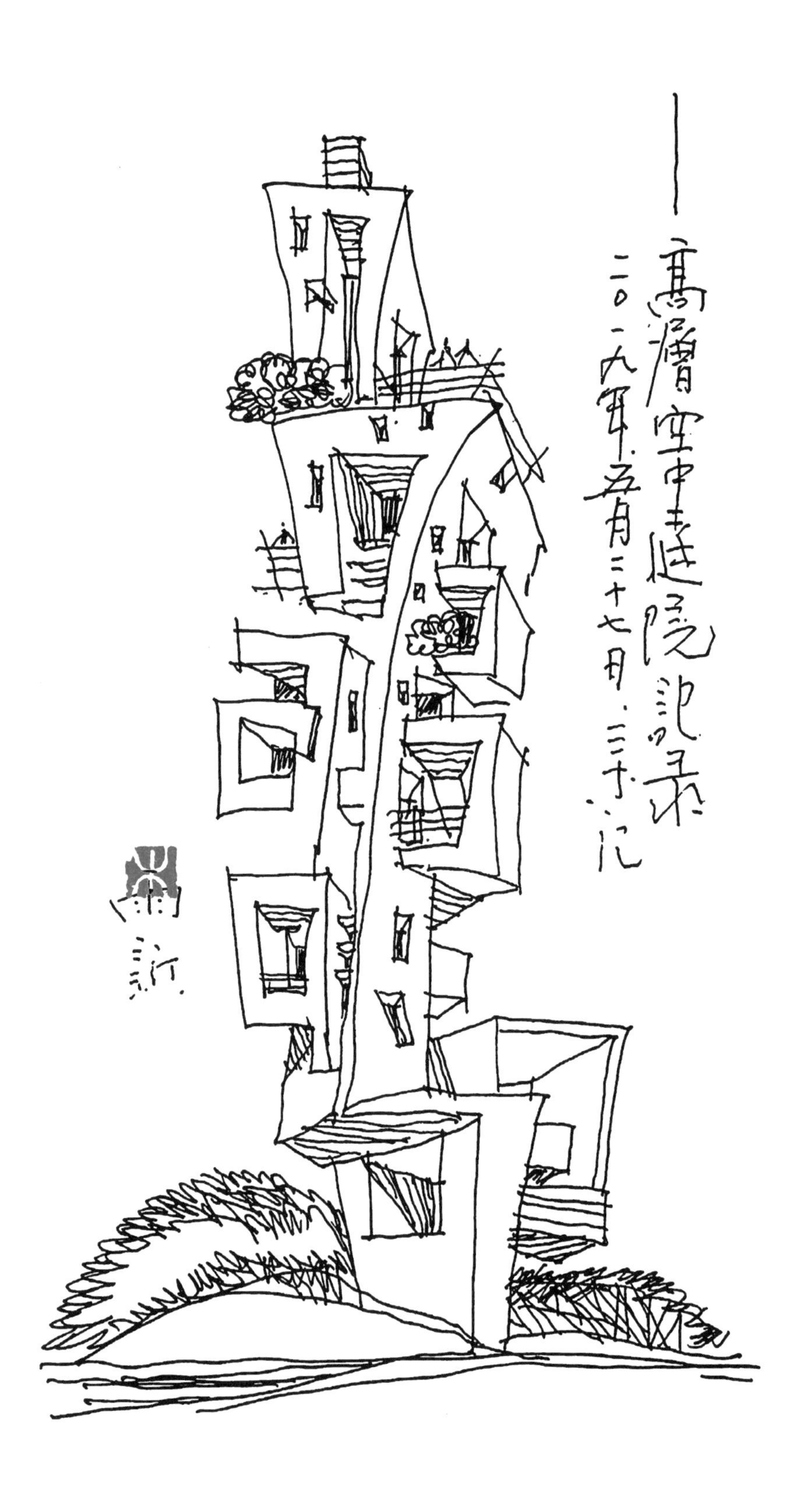
高層空中庭院記錄
二〇一九年五月二十七日

特色小镇

鄉村建設.
2019.3.27.

2014.3.27
2014.3.27

佛心寺
2019.3.23

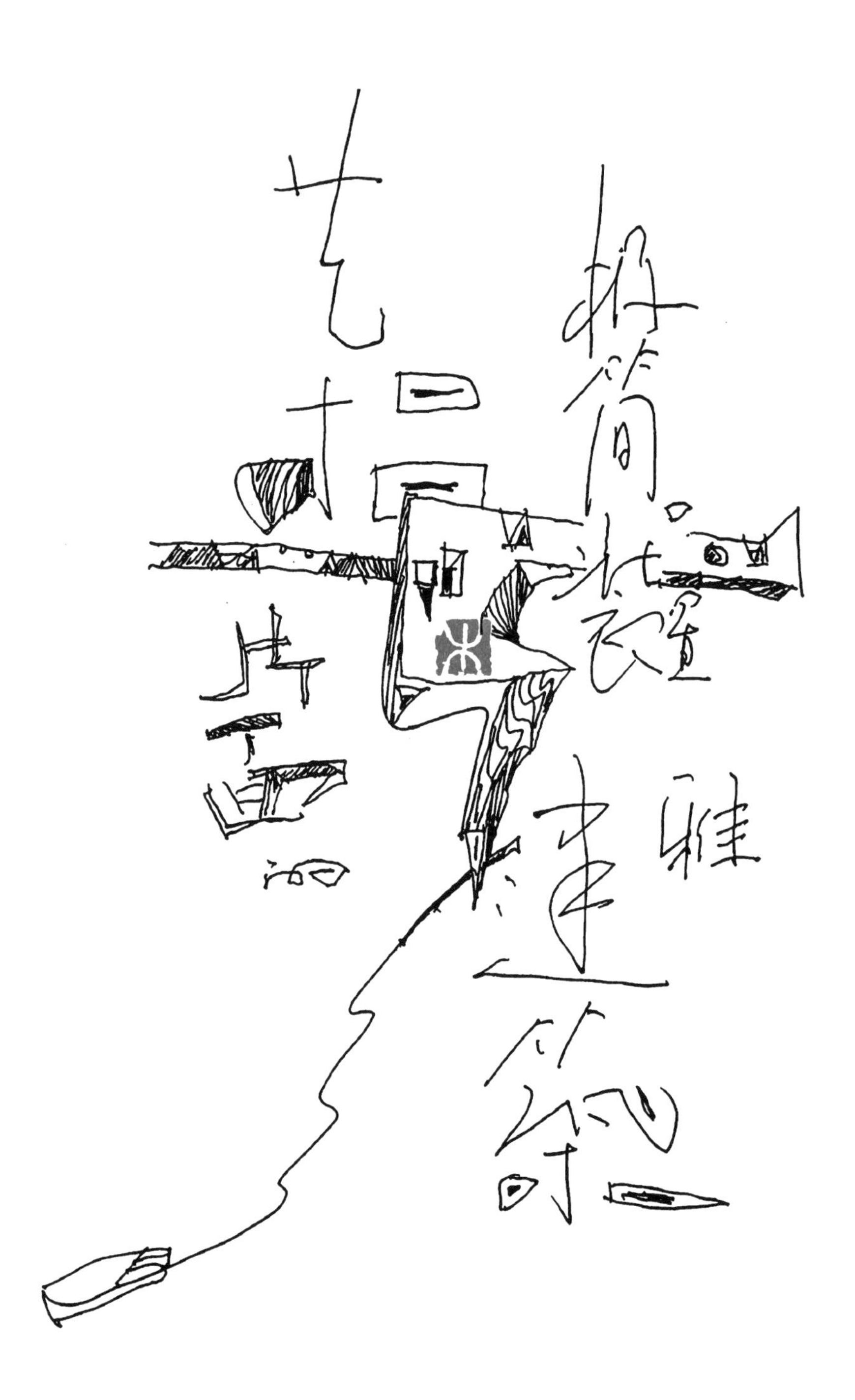

Boutique.
mote
I
DO

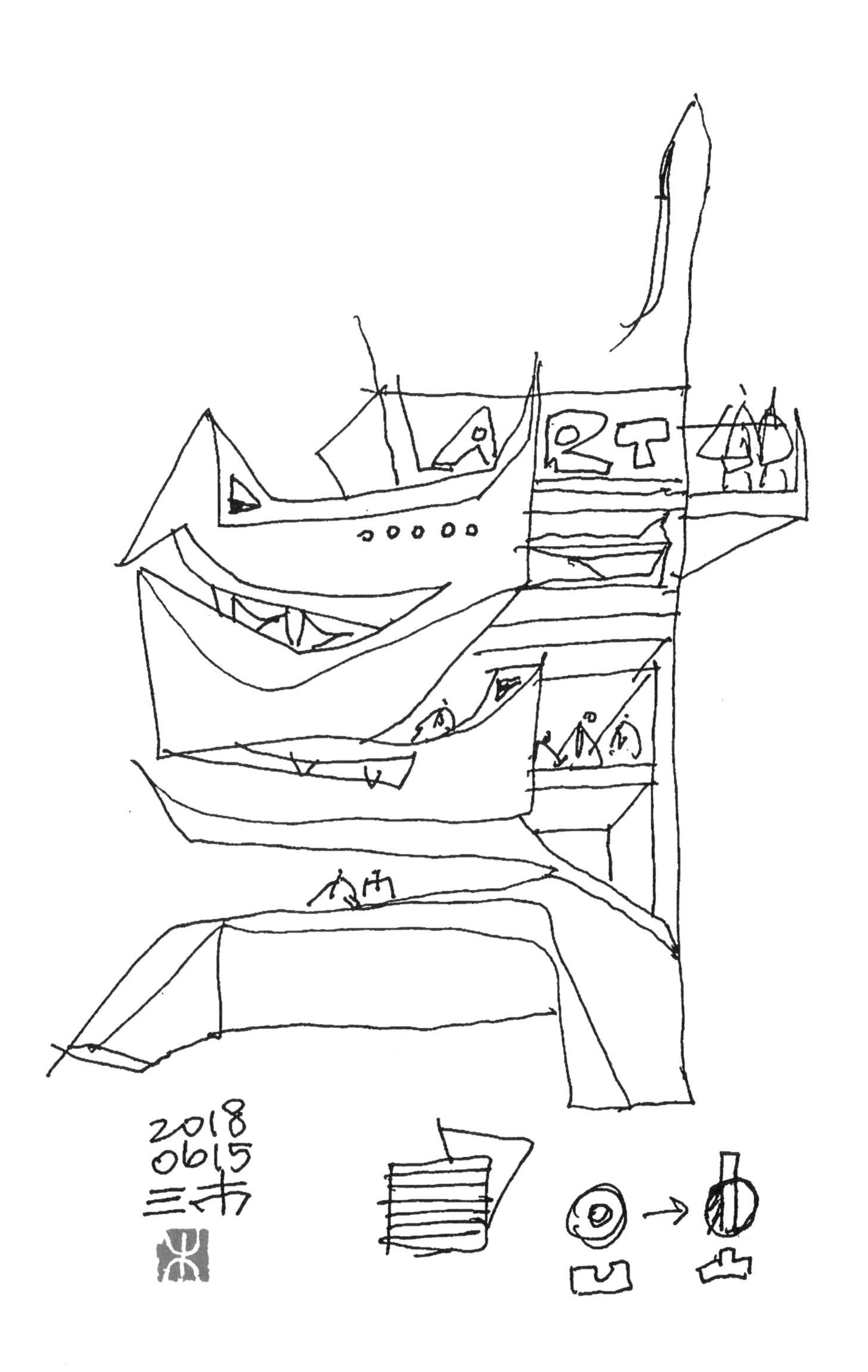
ART
2018
0615
三书

游.園.
2018
0615

2018.06.15

2018.06.15

日与水
2018.06.15

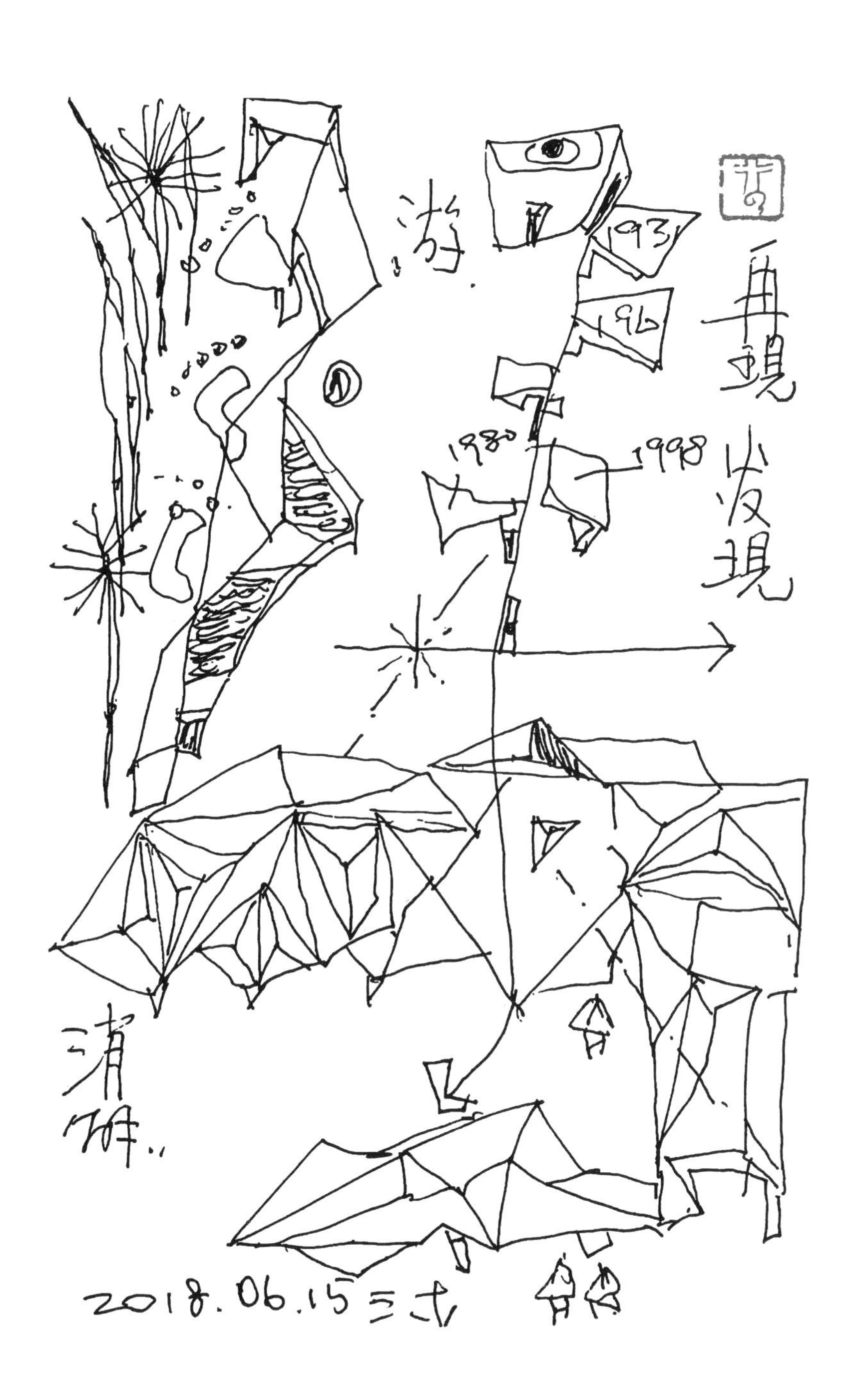

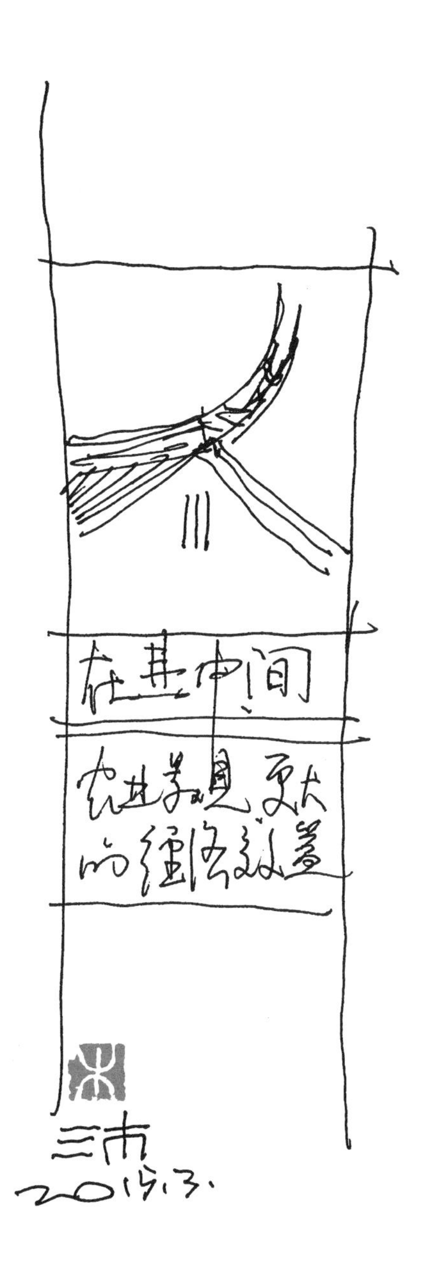
在其中间

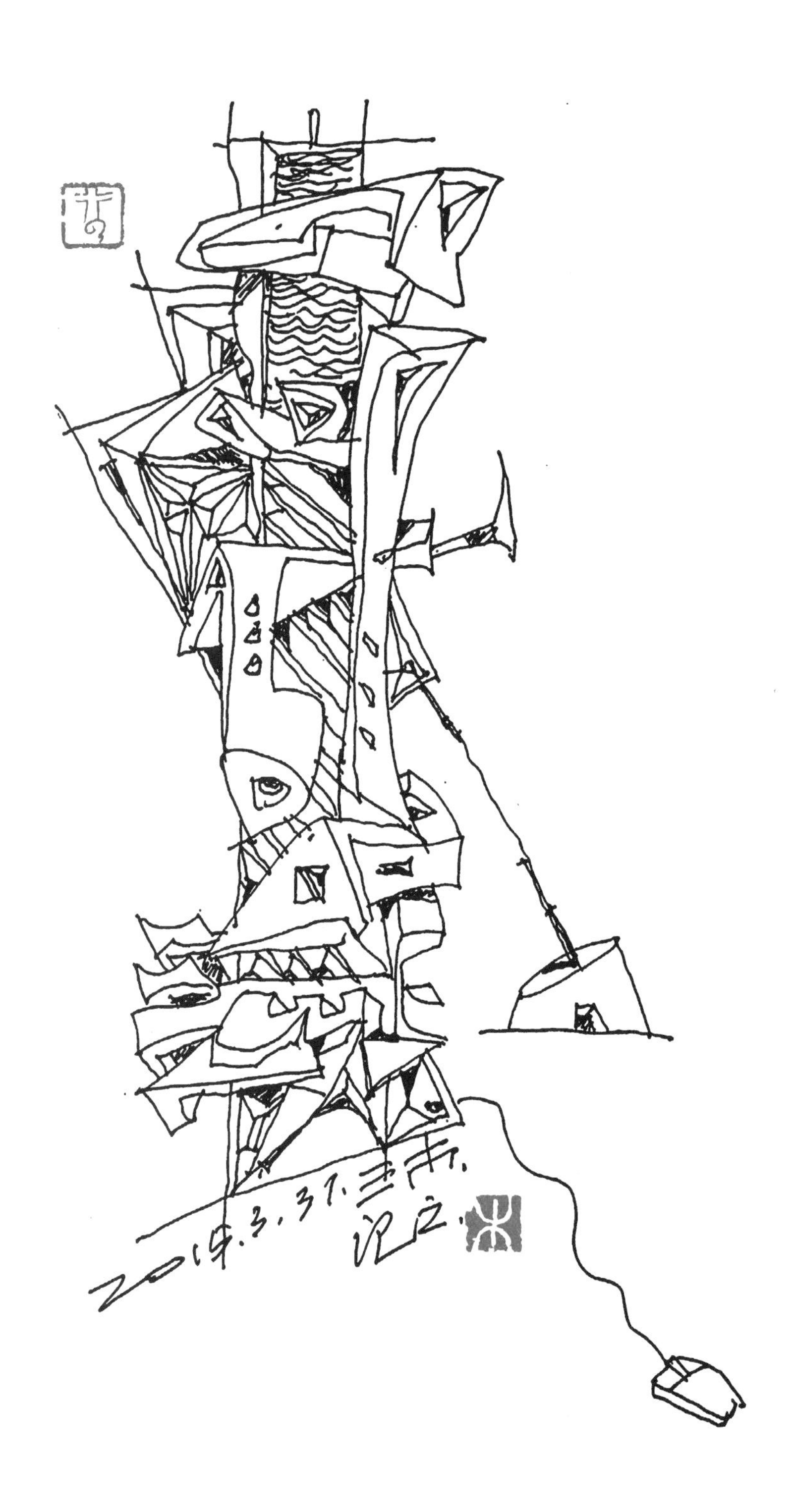
2014.3.31.

AB

2015.3.31

2015.3.31

温泉河
2015.3.31

Lin.
Yi
2015.3.31

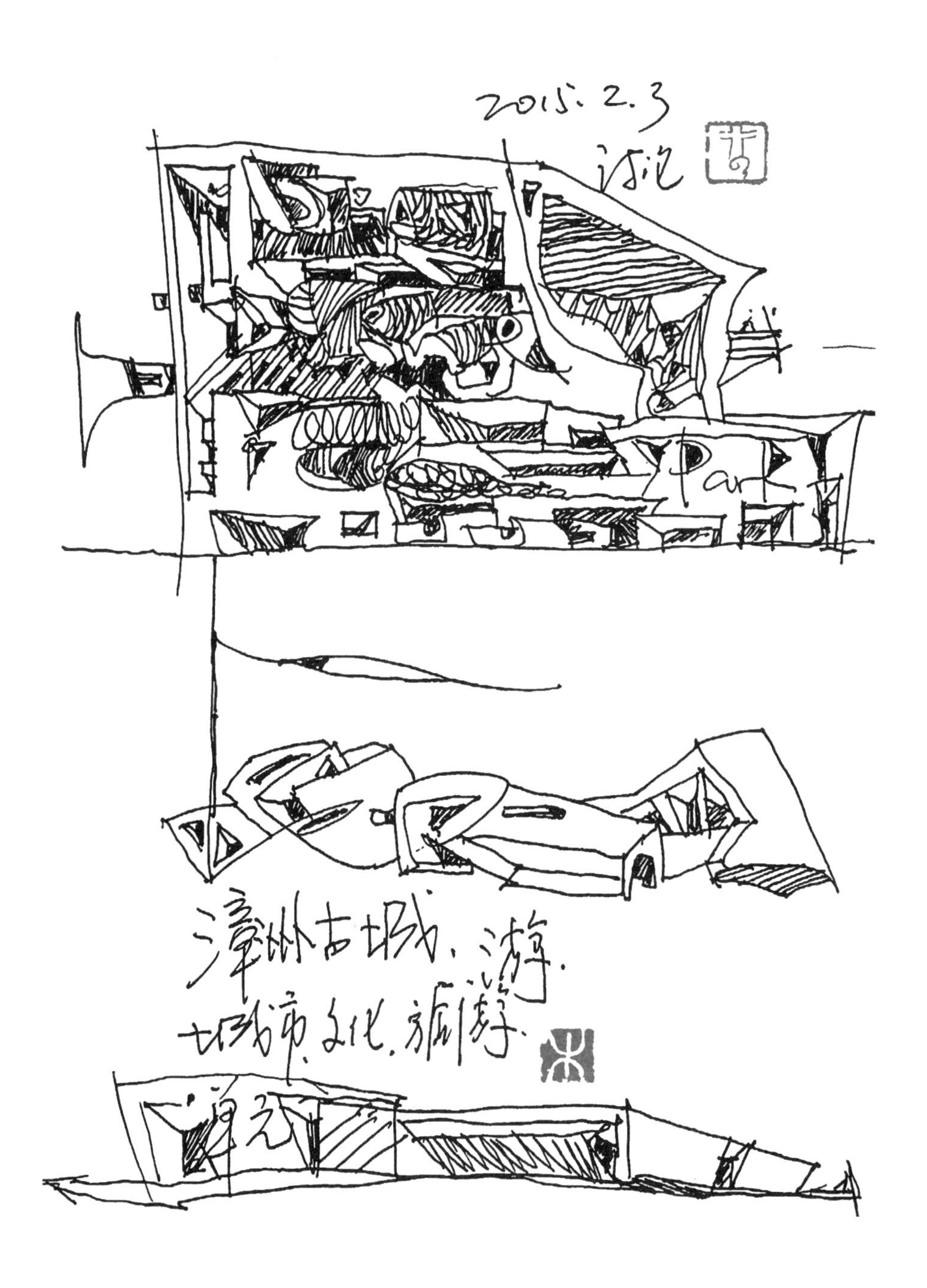
2015.2.3
Park
漳州古城
城市文化、旅游

沙坡尾
2015.1.21

2015.1.21

空間
2015.1.21

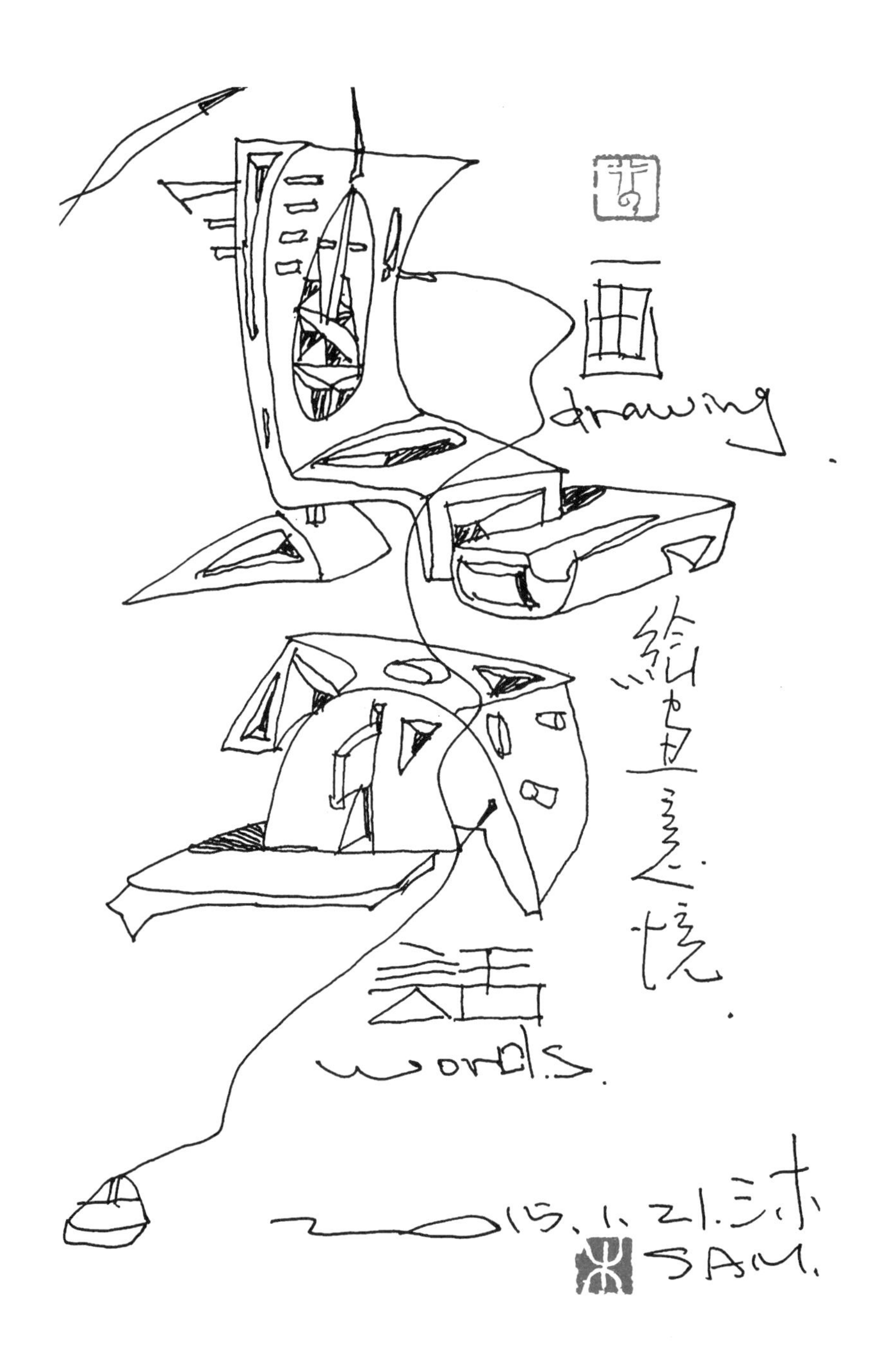
drawing.
words.
SAM.

泉州植物园.
2015.
1.
28.

水
WATER.
2015.1.24.

2015.1.23

濱水空间.
2015.1.21
XMU-SAM

Rain
Earth Building
2015.1.18

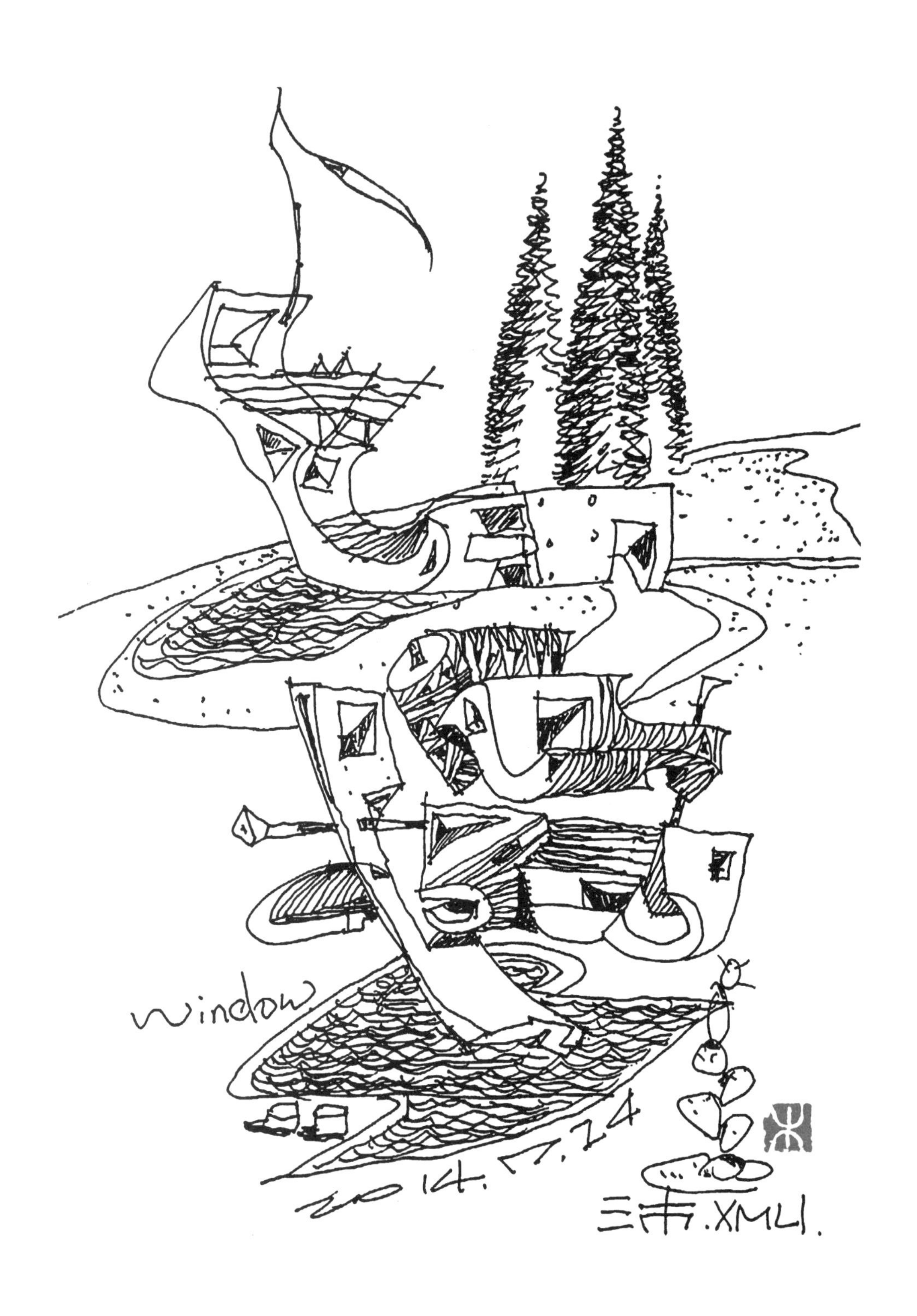
Window
2014.7.24
三市.XMLI.

闽南
2014.7.23
XMU.

交通
2014.7.24

2014.7.23

2014.7.23

2013.9.22

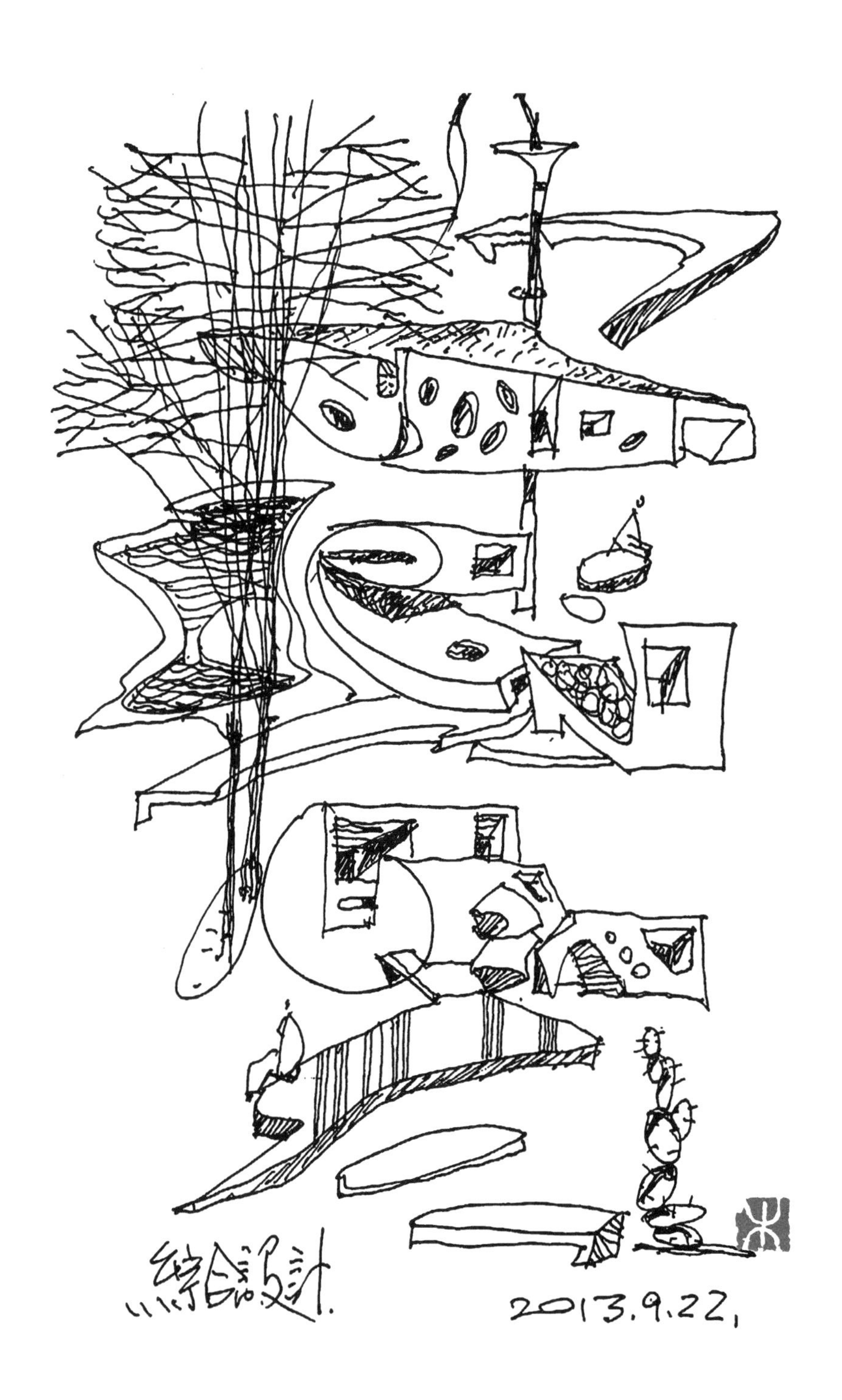
綜合設計.
2013.9.22,

ING
老人住宅.
2013.9.22

空間。
虚。
天街。
泰山。
2013.6

2013.3.11. 三市

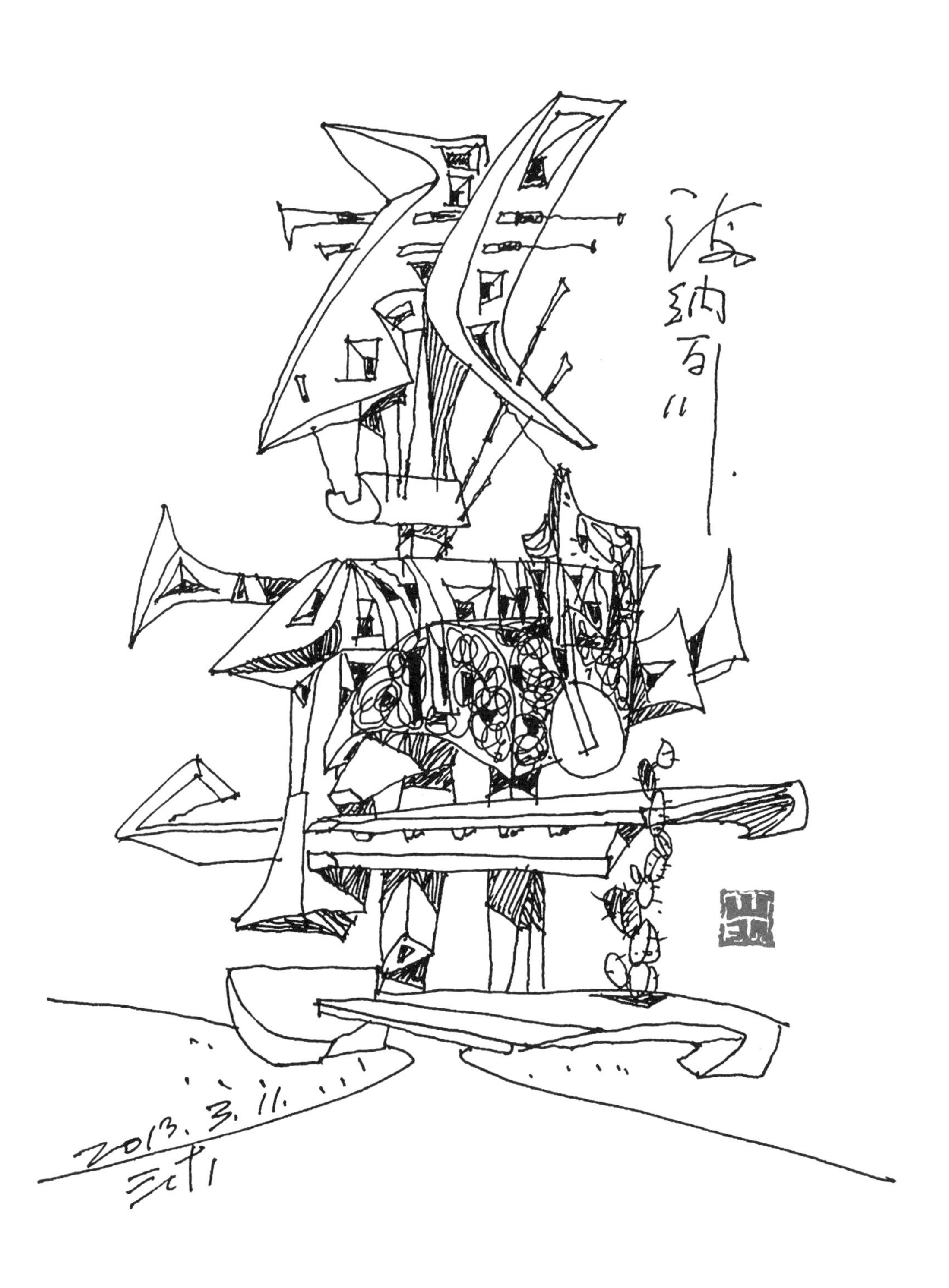
2013.3.11.

2013.3.11.

2013.3.11

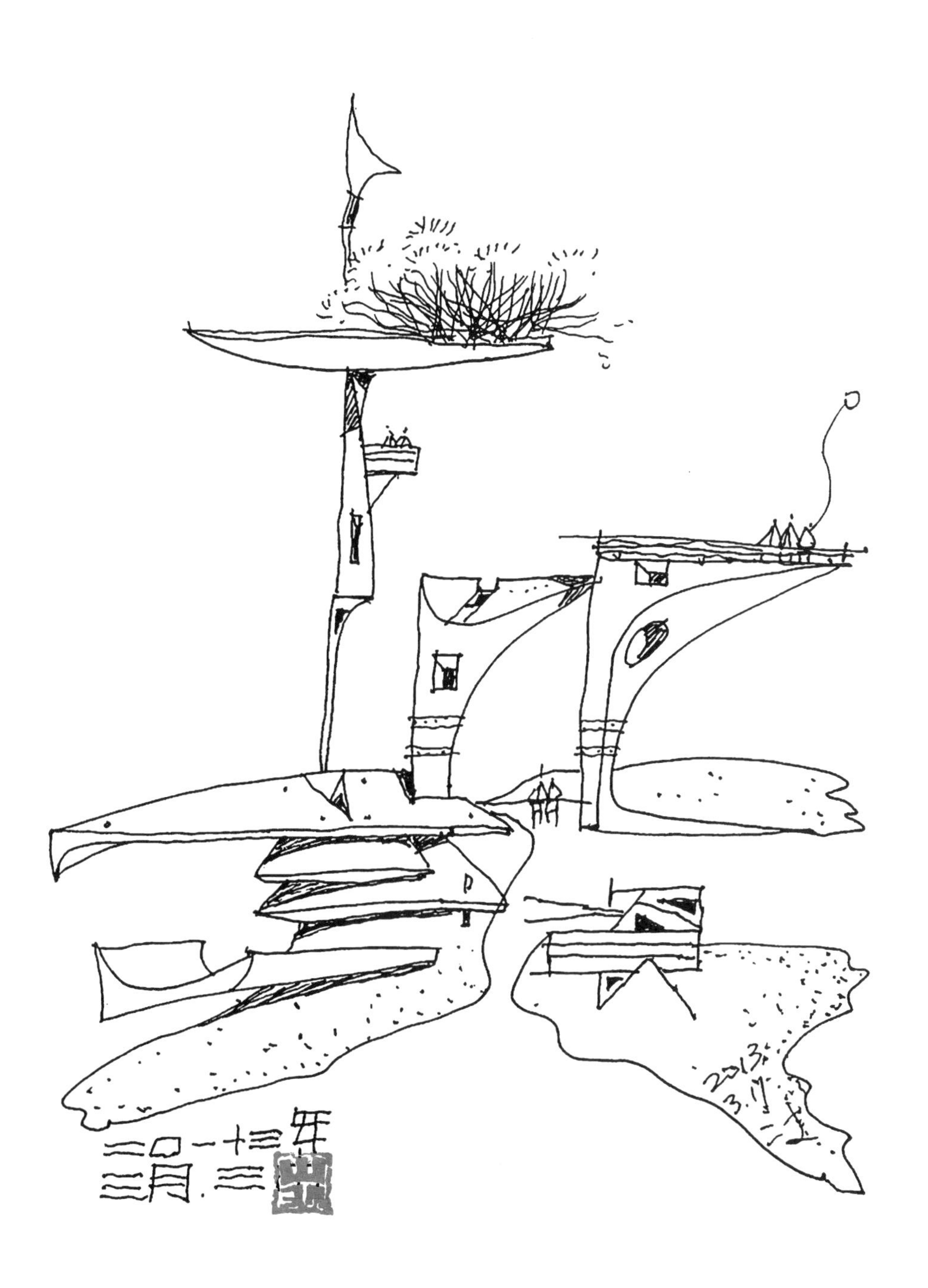

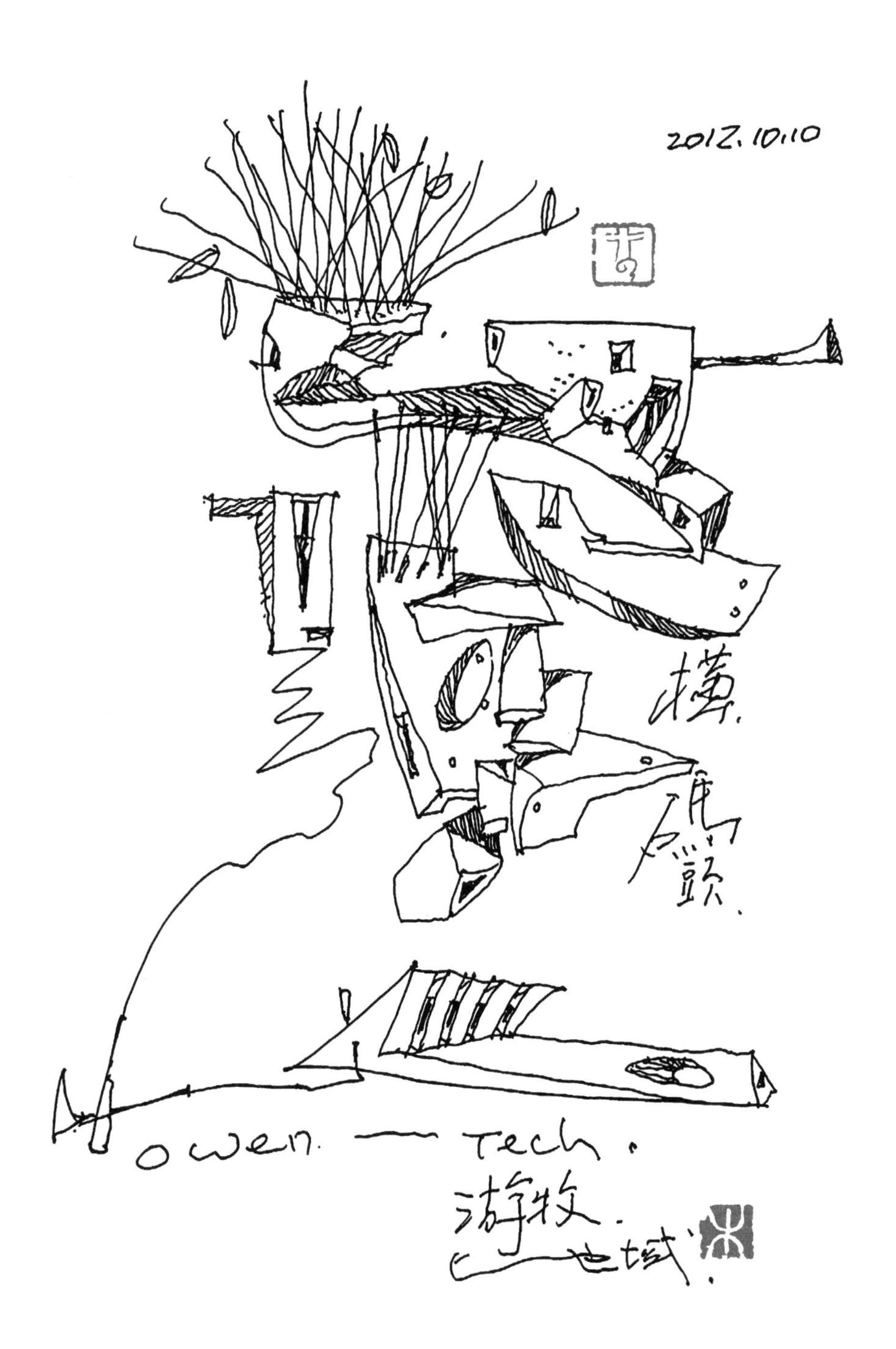
2012.10.10
owen — Tech.
游牧

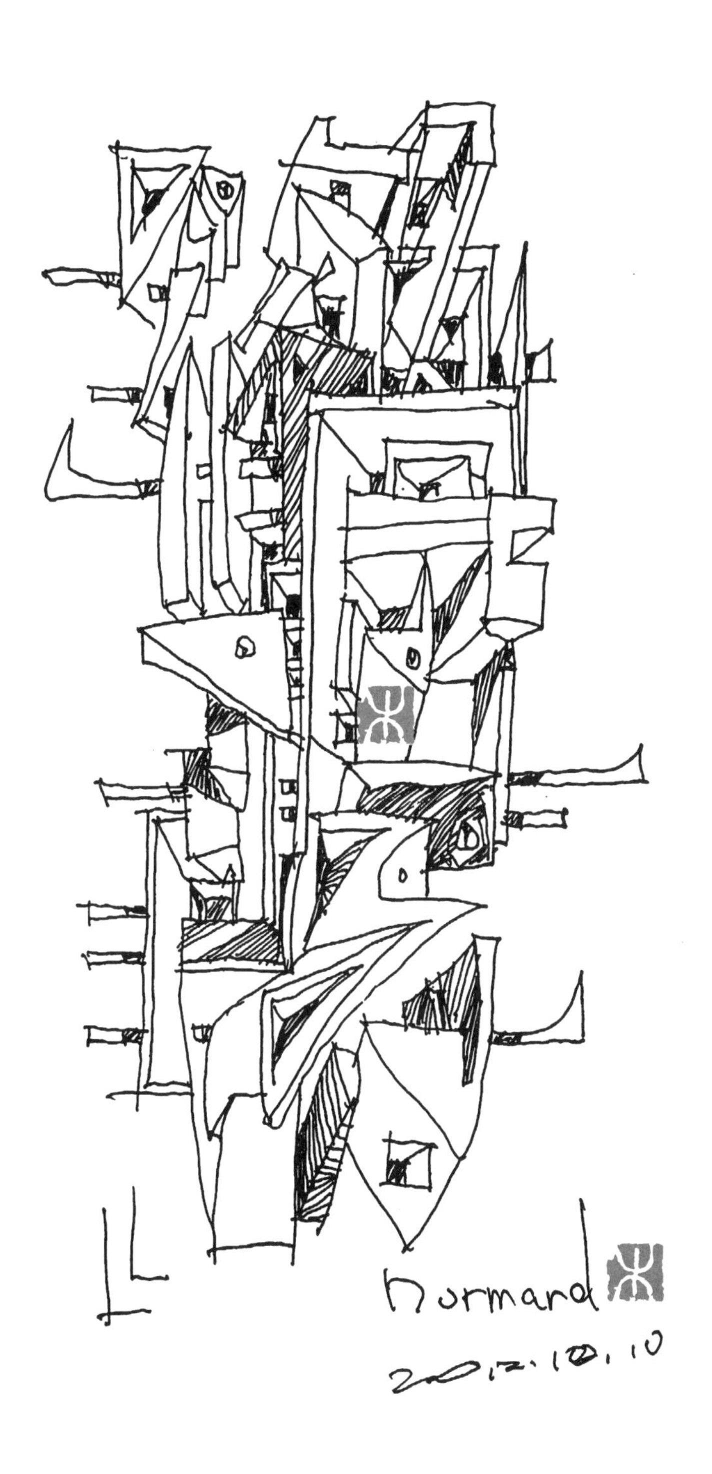
Normand
2012.10.10

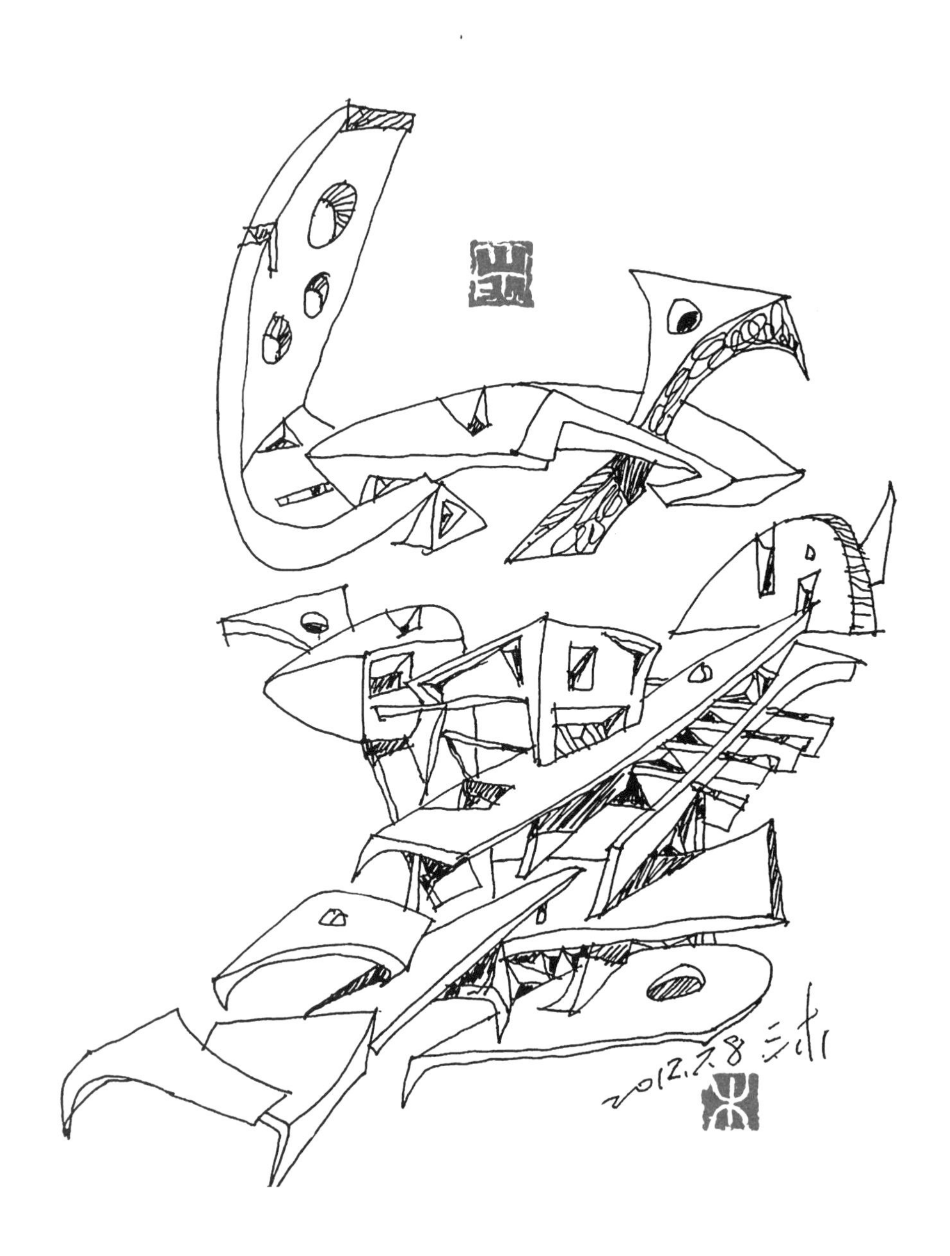

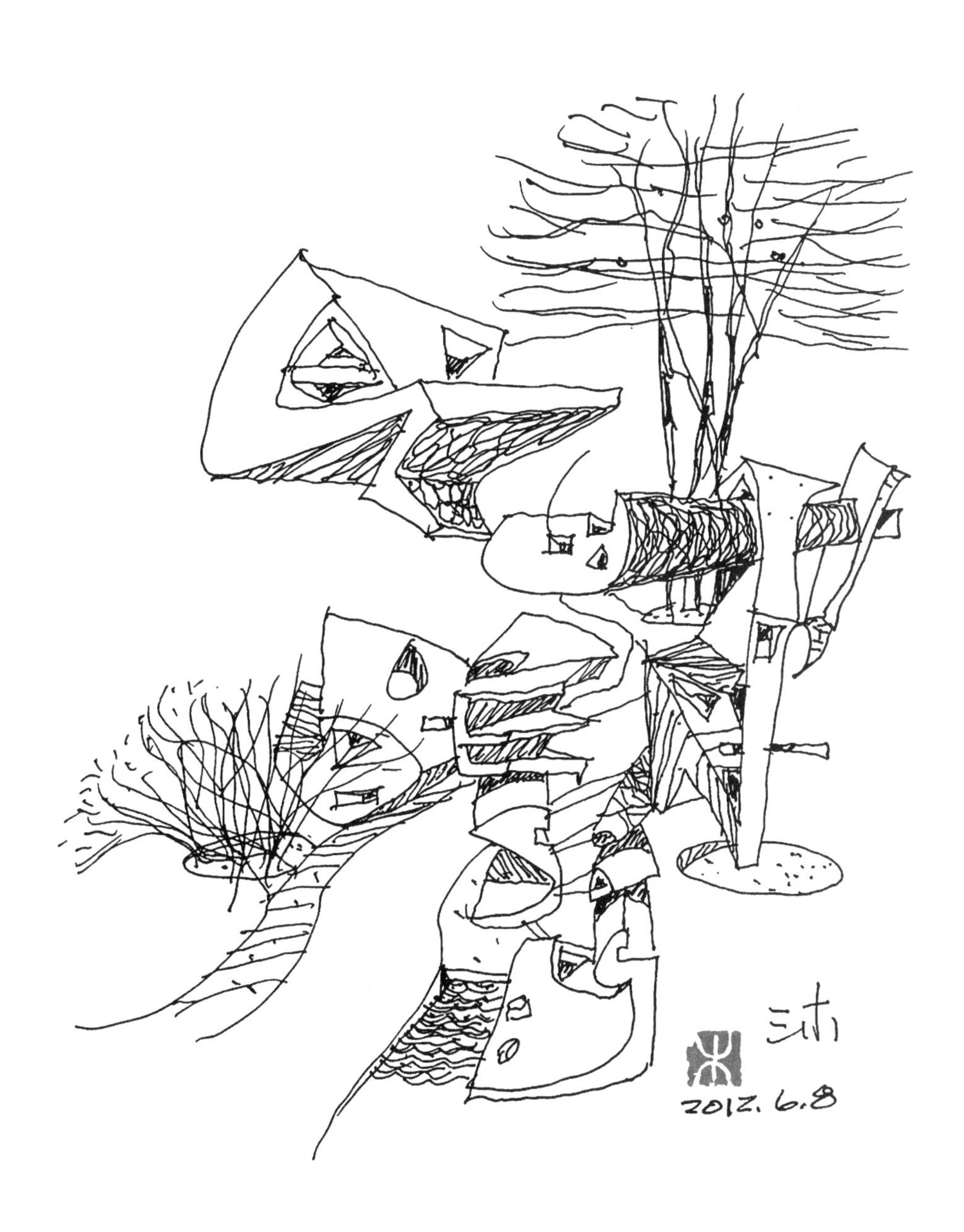
2012.6.8

三木
2012.6.5

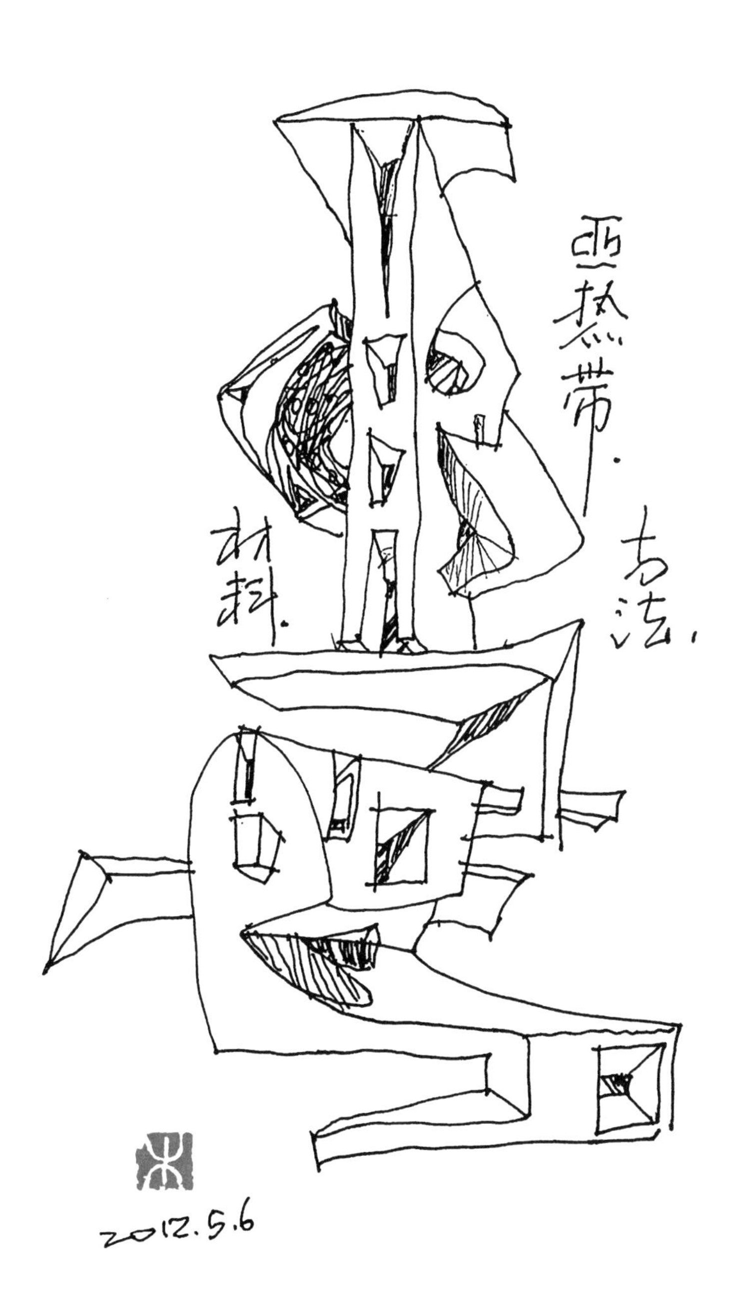
亚热带.
方法,
材料.
2012.5.6

中山路
2012.5.6

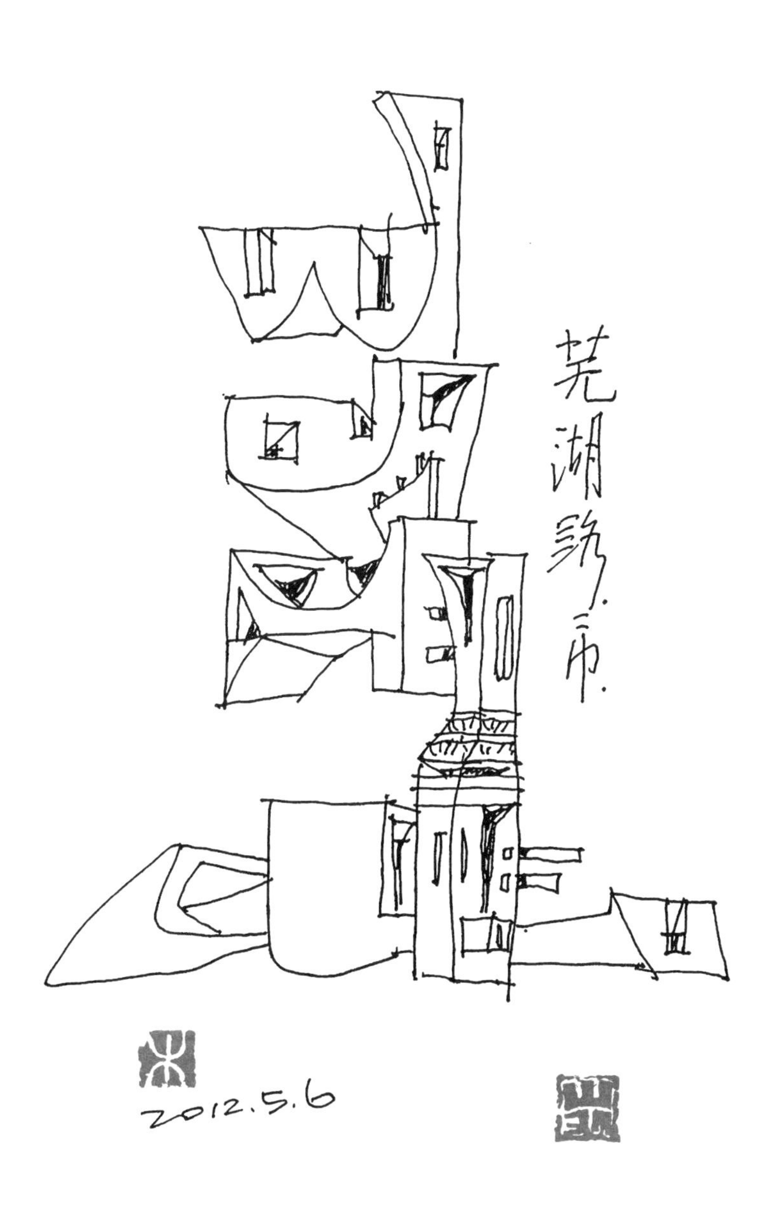
2012.5.6

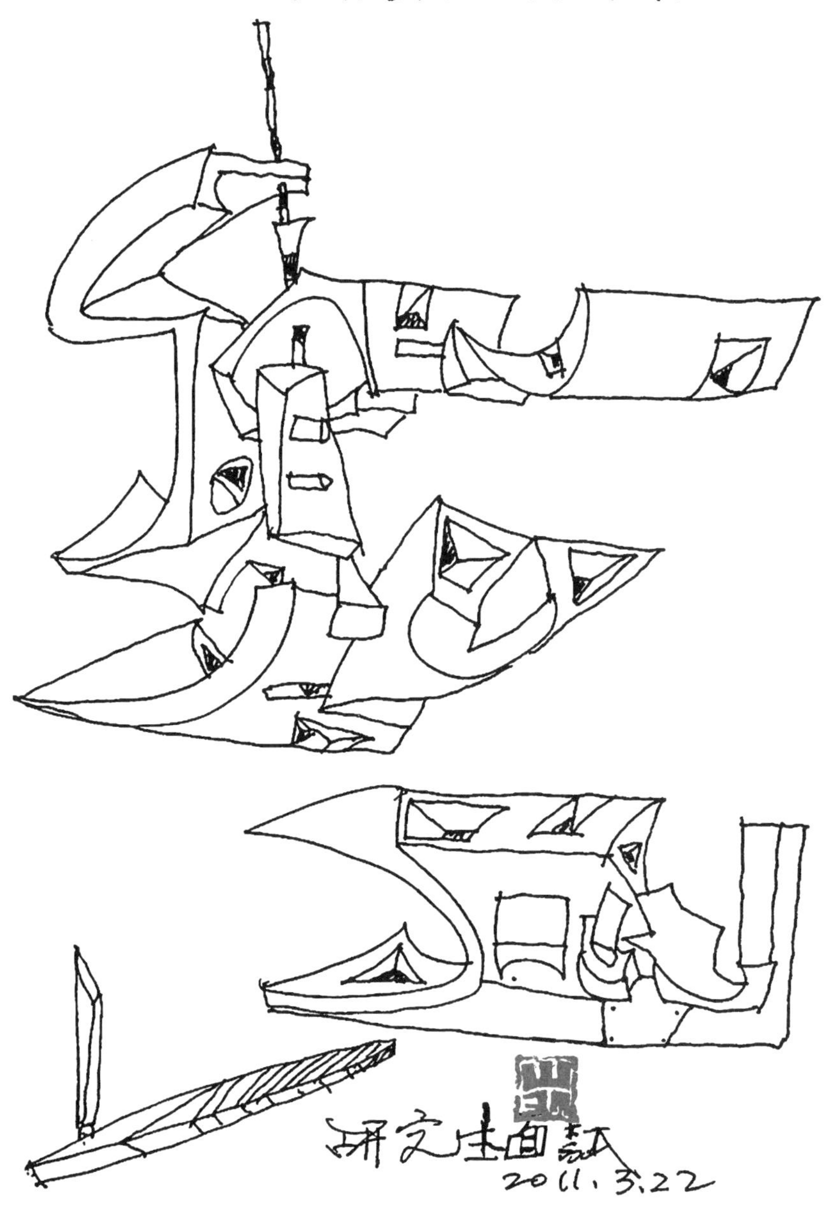
2011.3.22. XMH.
研究生自款
2011.3.22

XMU.三市.
2012.4

2012.4.5

2011.4.12

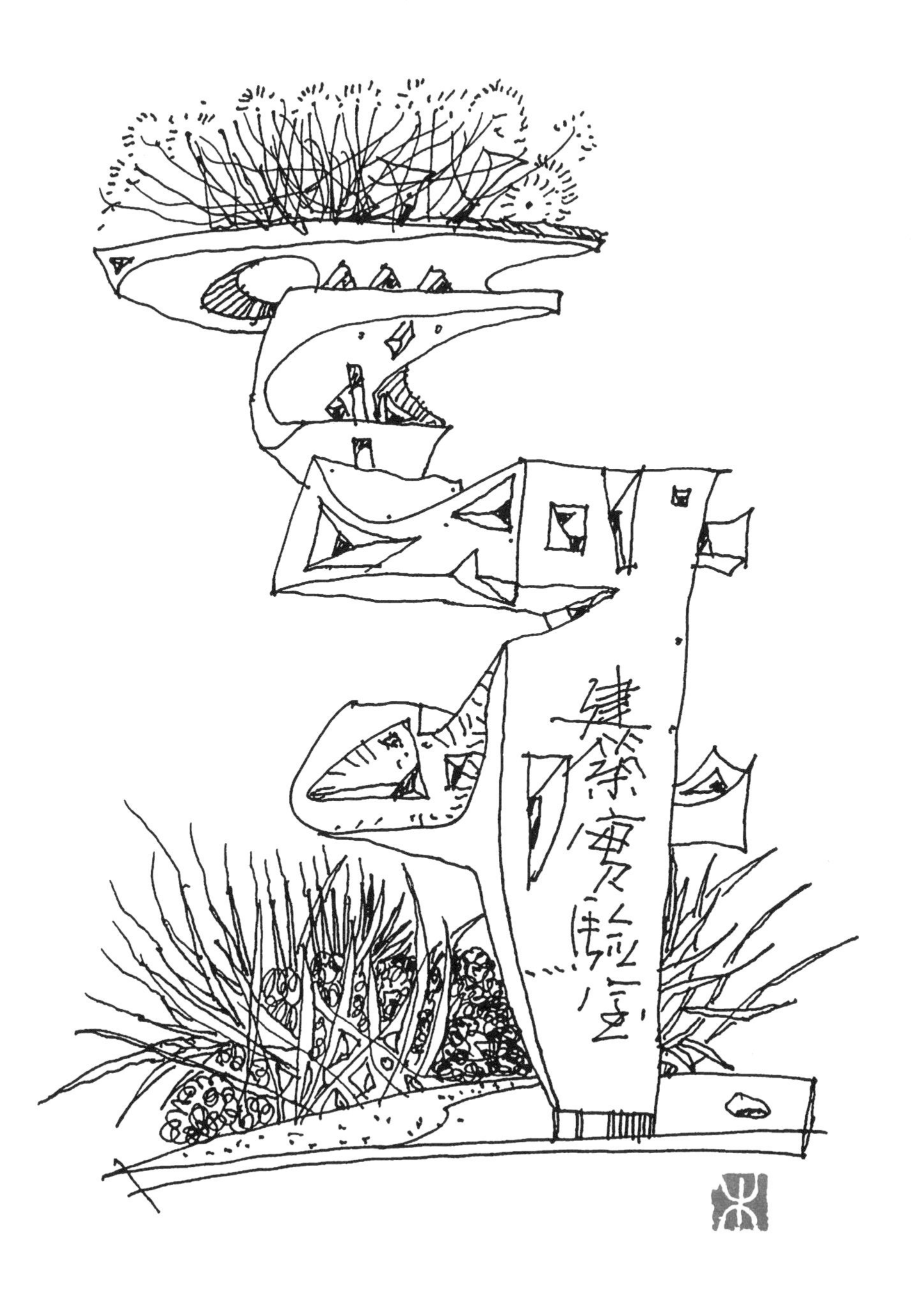

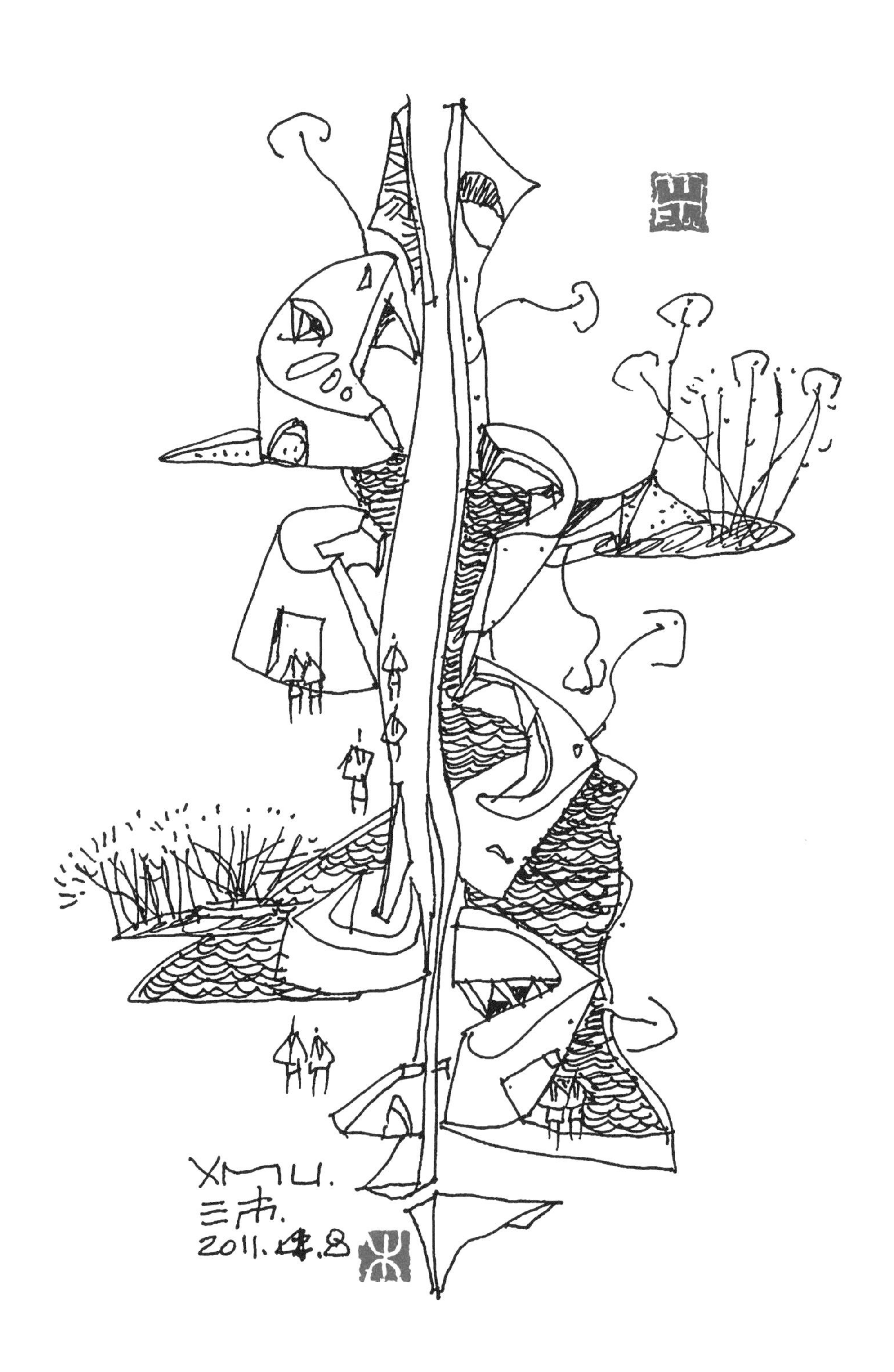
XMU.
三市.
2011.4.8

2019.7.25.

2011. 3.25

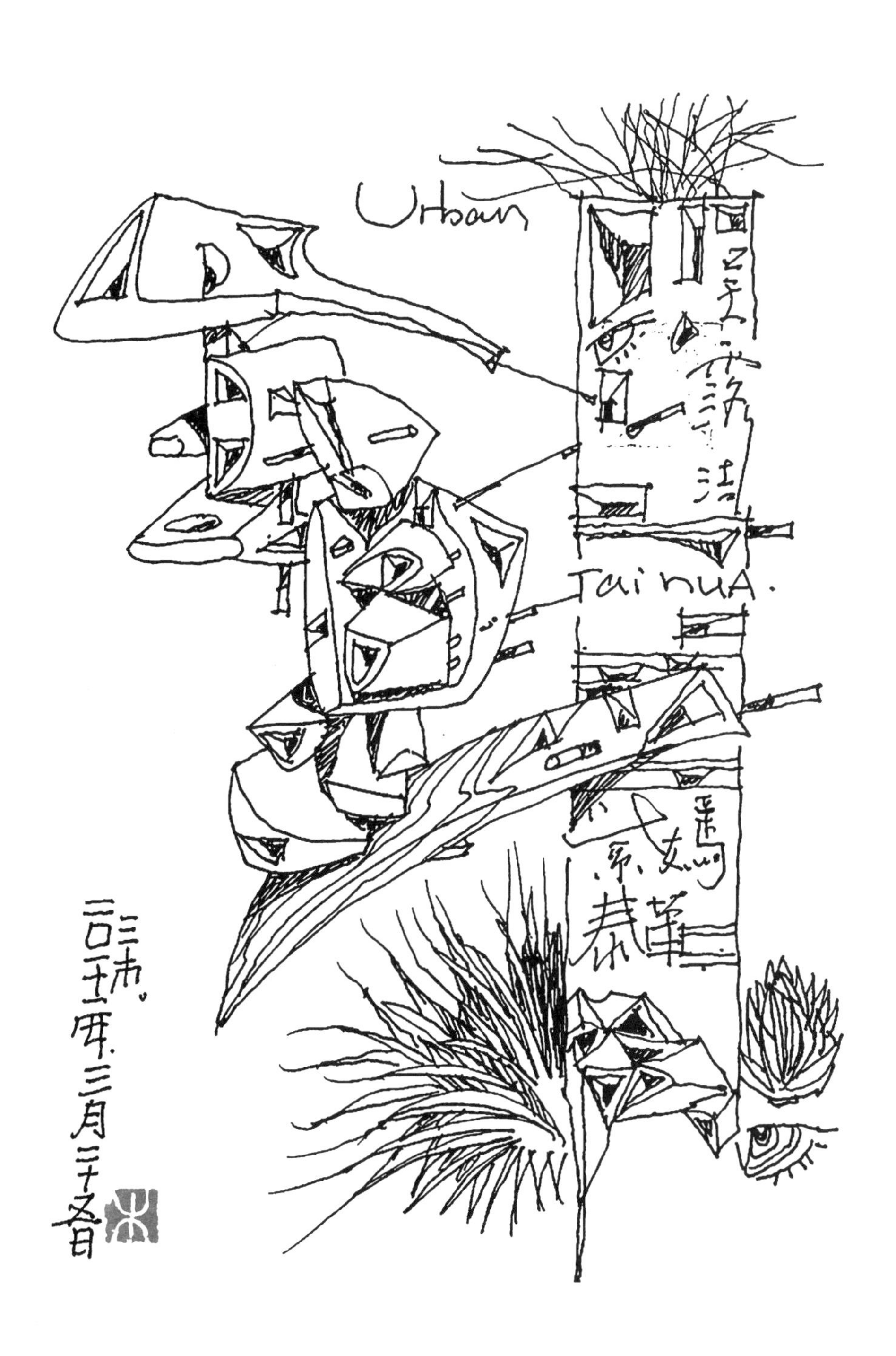
Urban
Tai nuA.

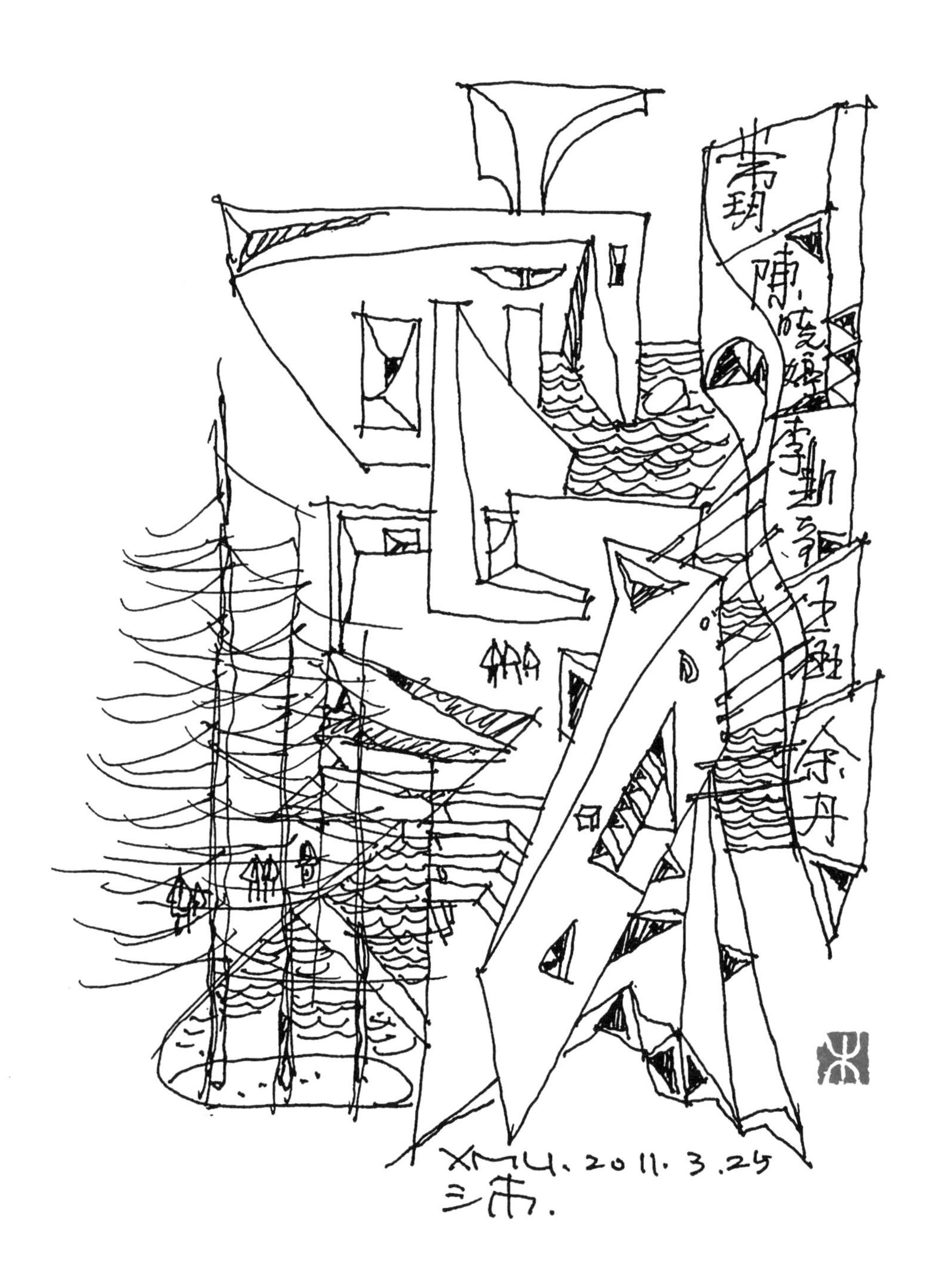
XMU. 2011. 3.25

技术.
ArchitectURE
建筑.

1. 2011.3.25
5.

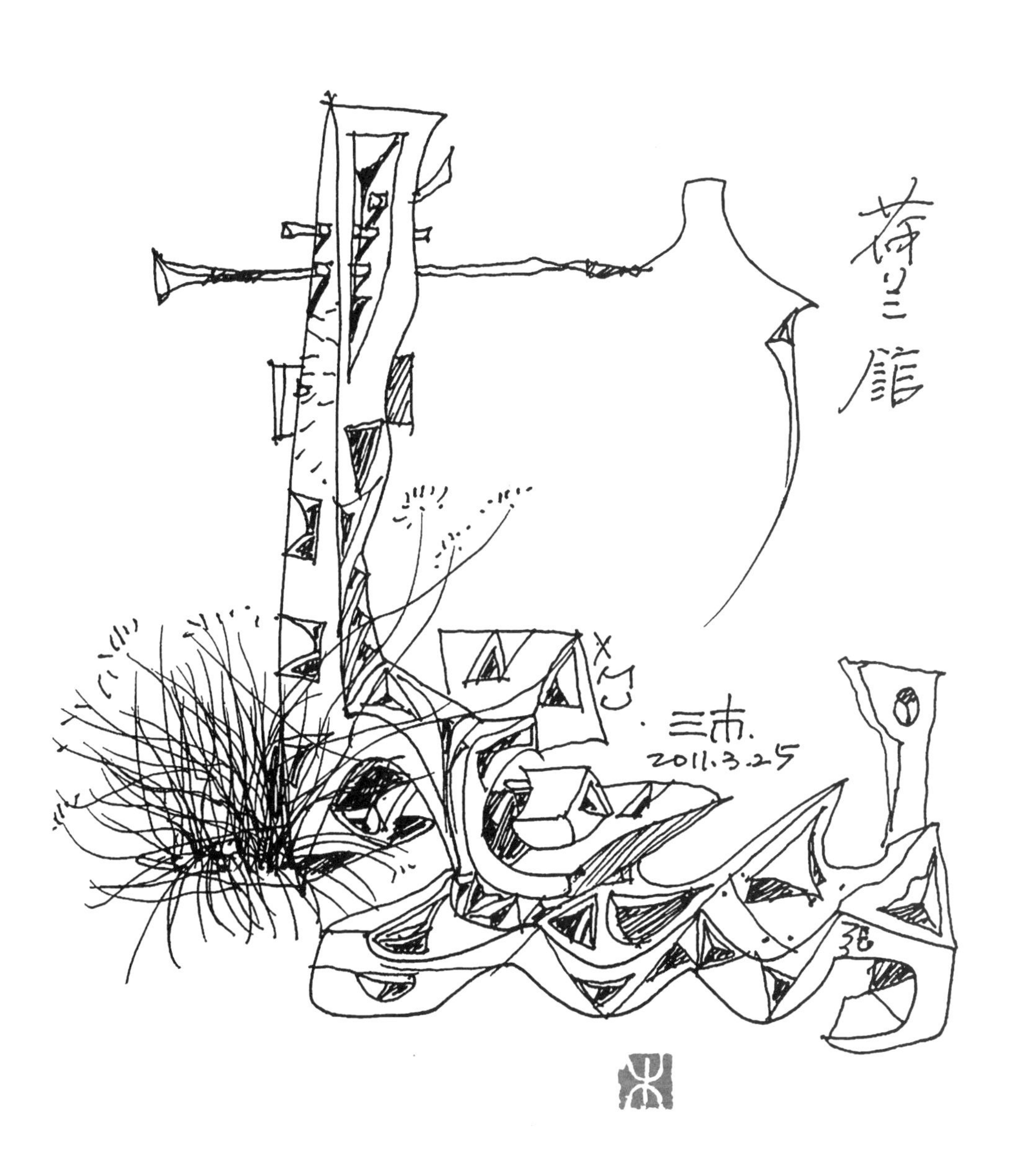
荷兰馆
2011.3.25

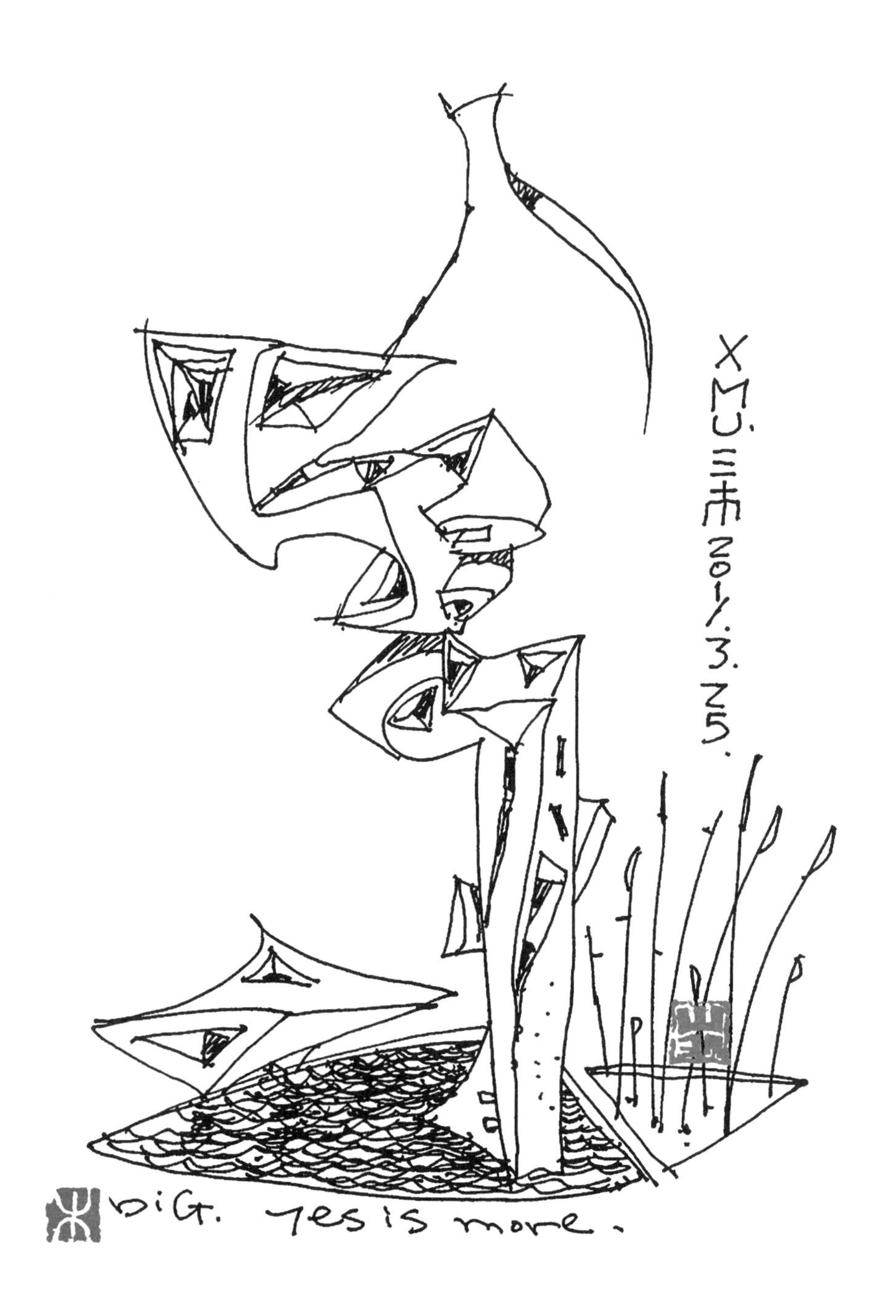
XMU 三木 2011.3.25.
biG. yes is more.

XMU.三石
2011. 3-25

藤本壮介
2011. 3. 25.

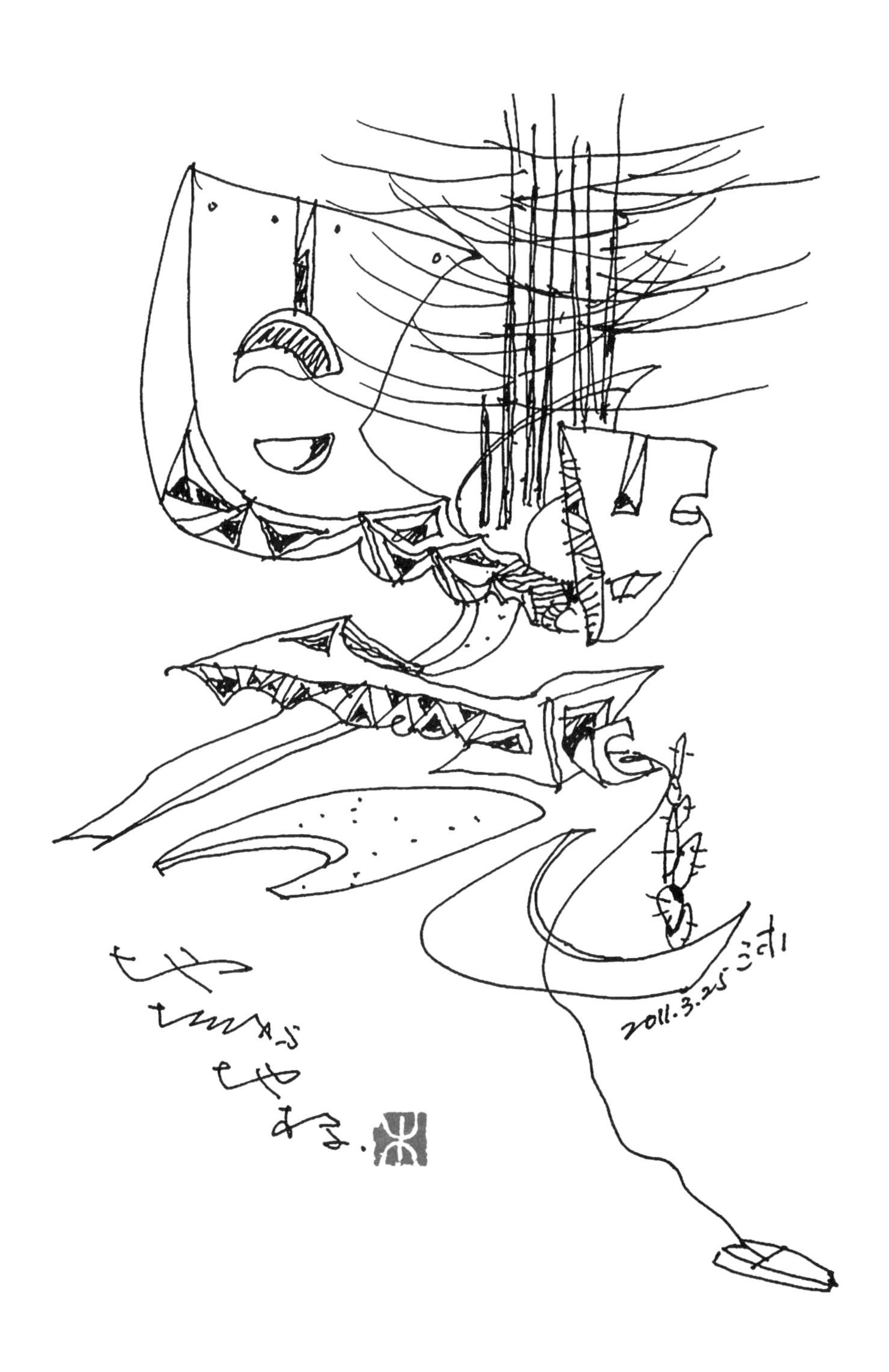

表皮
skin
2011.3.25.

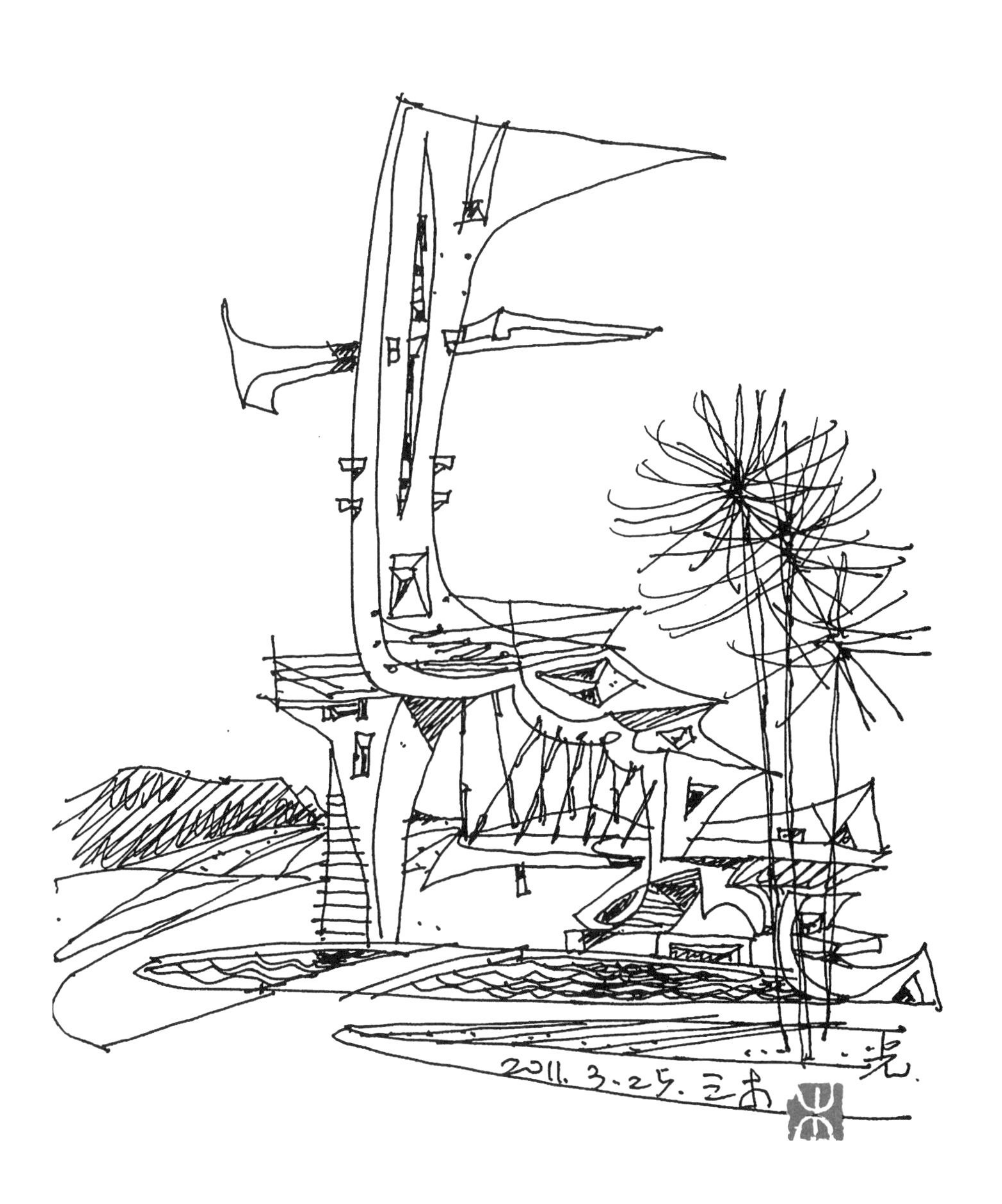
2011. 3. 25.

2013.3.22

2011.3.22. 三木. XMU

2011.3.22

2012.3.22

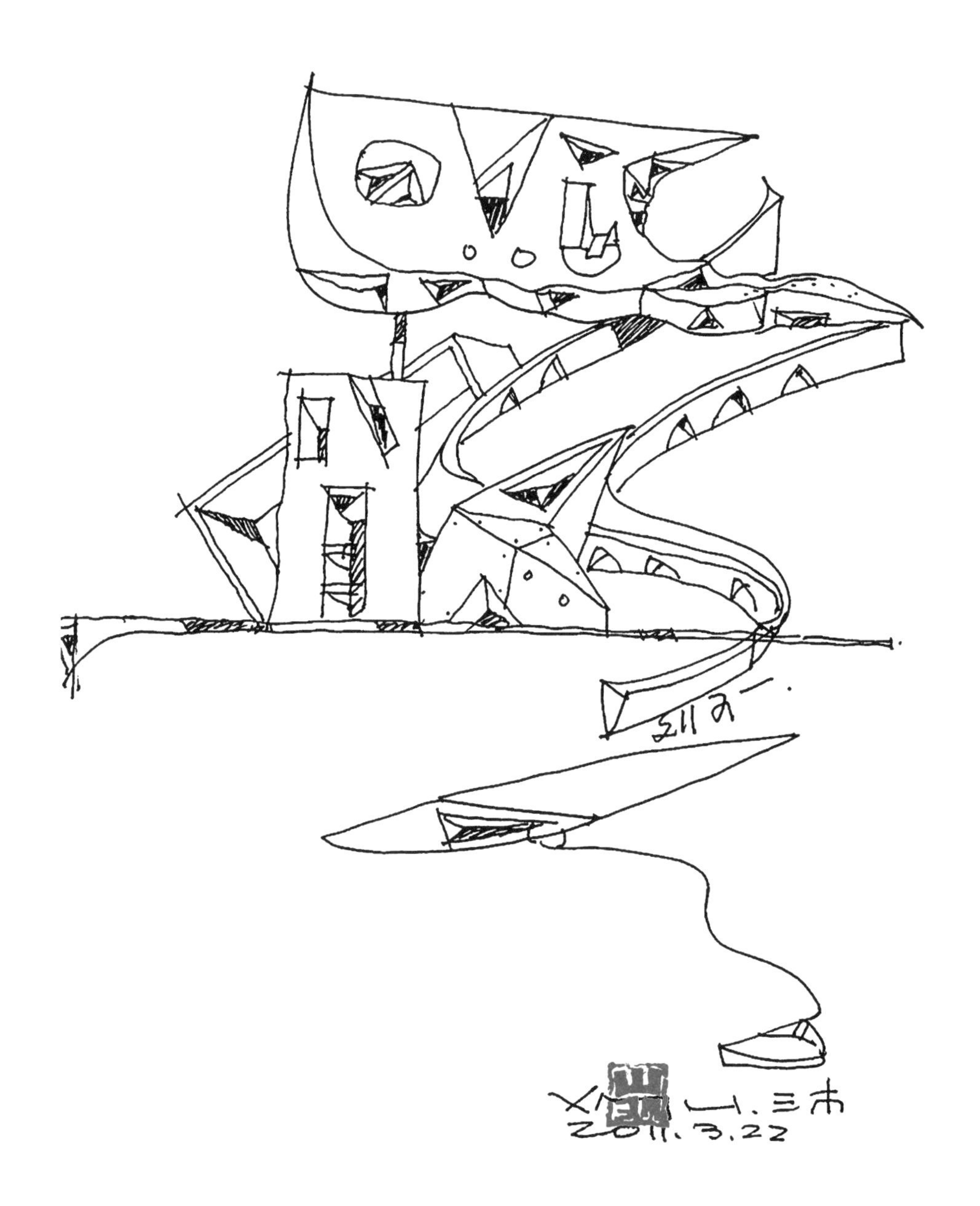

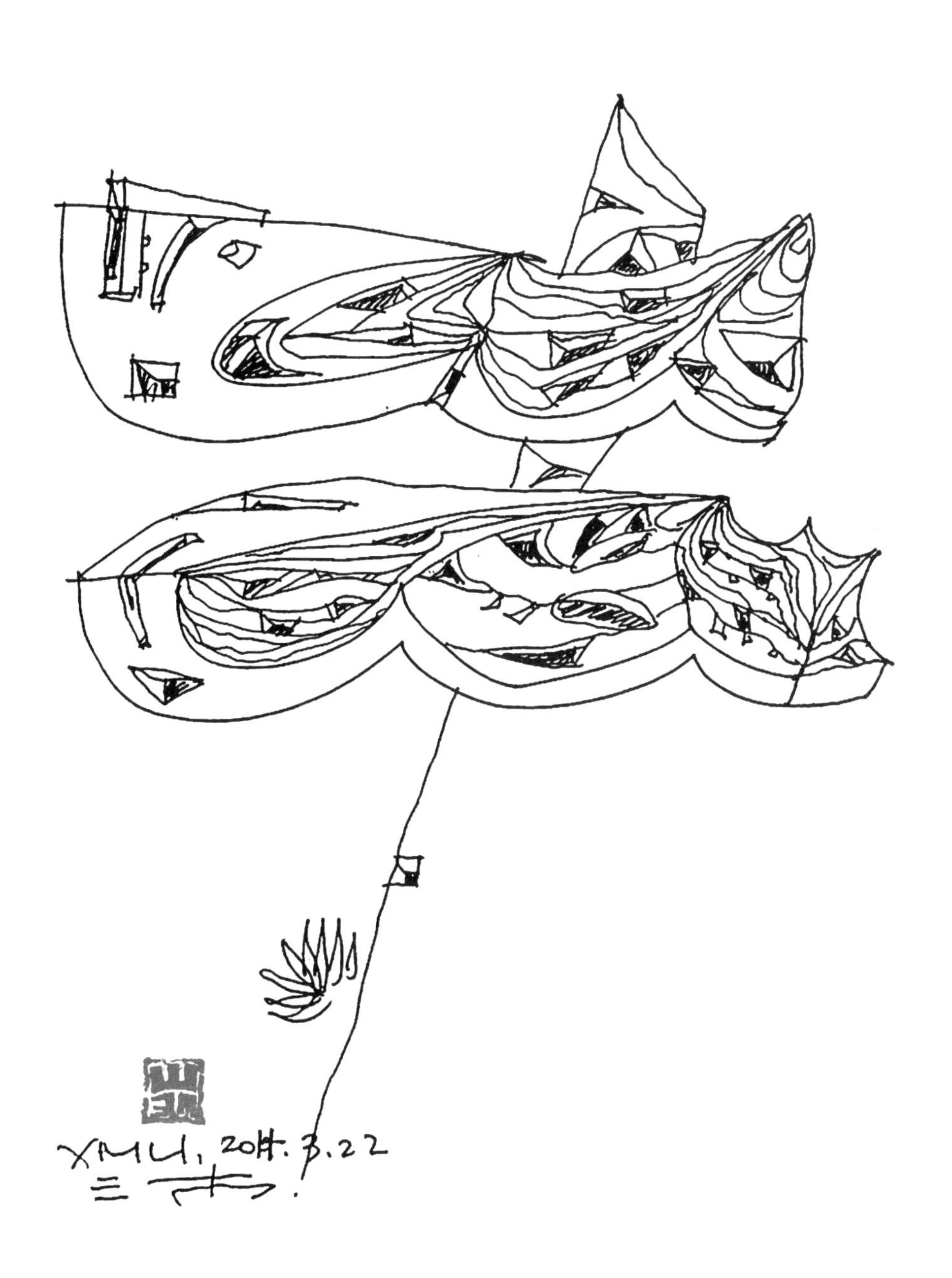
XMH, 2014.3.22

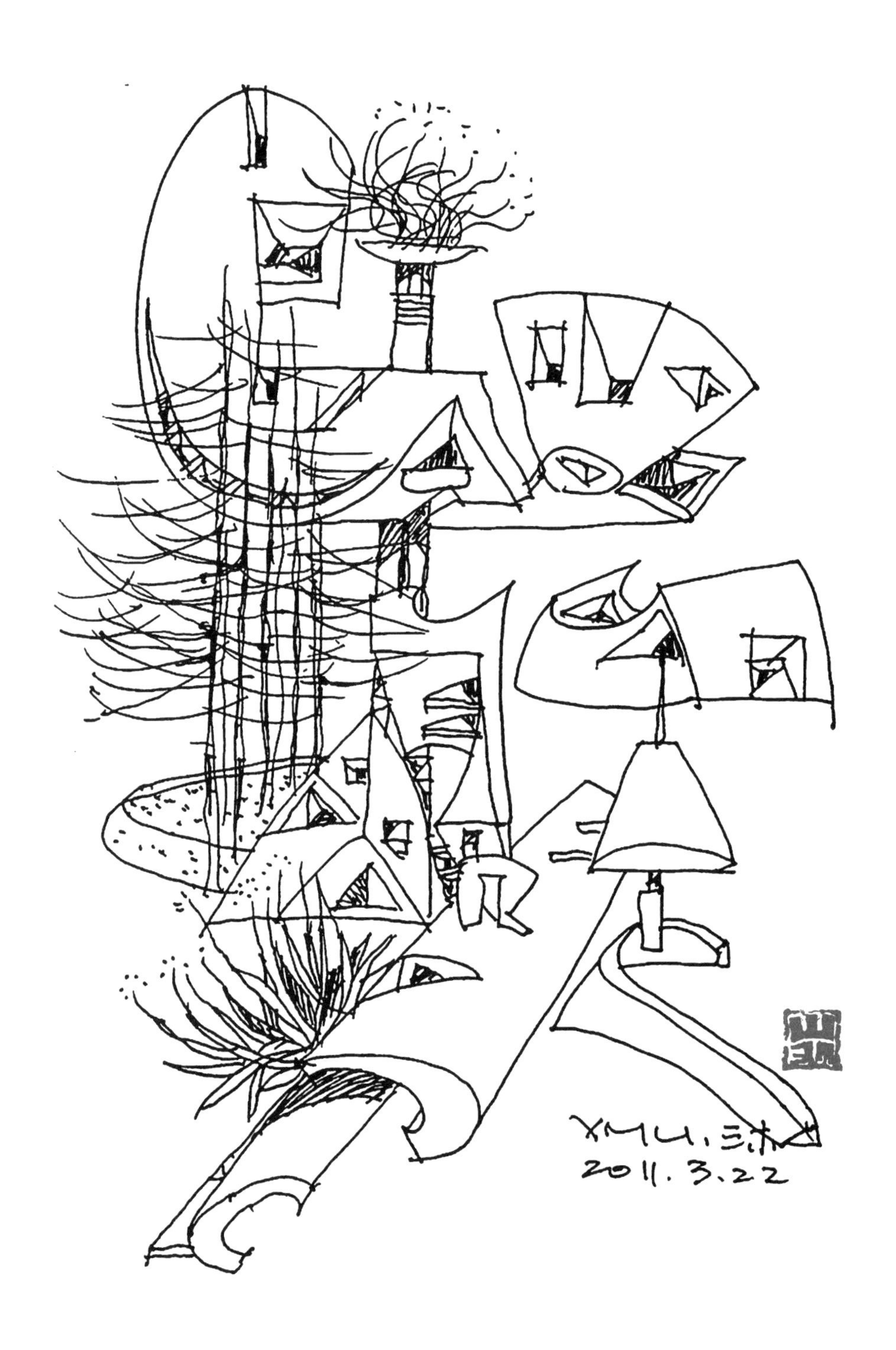
2011.3.22

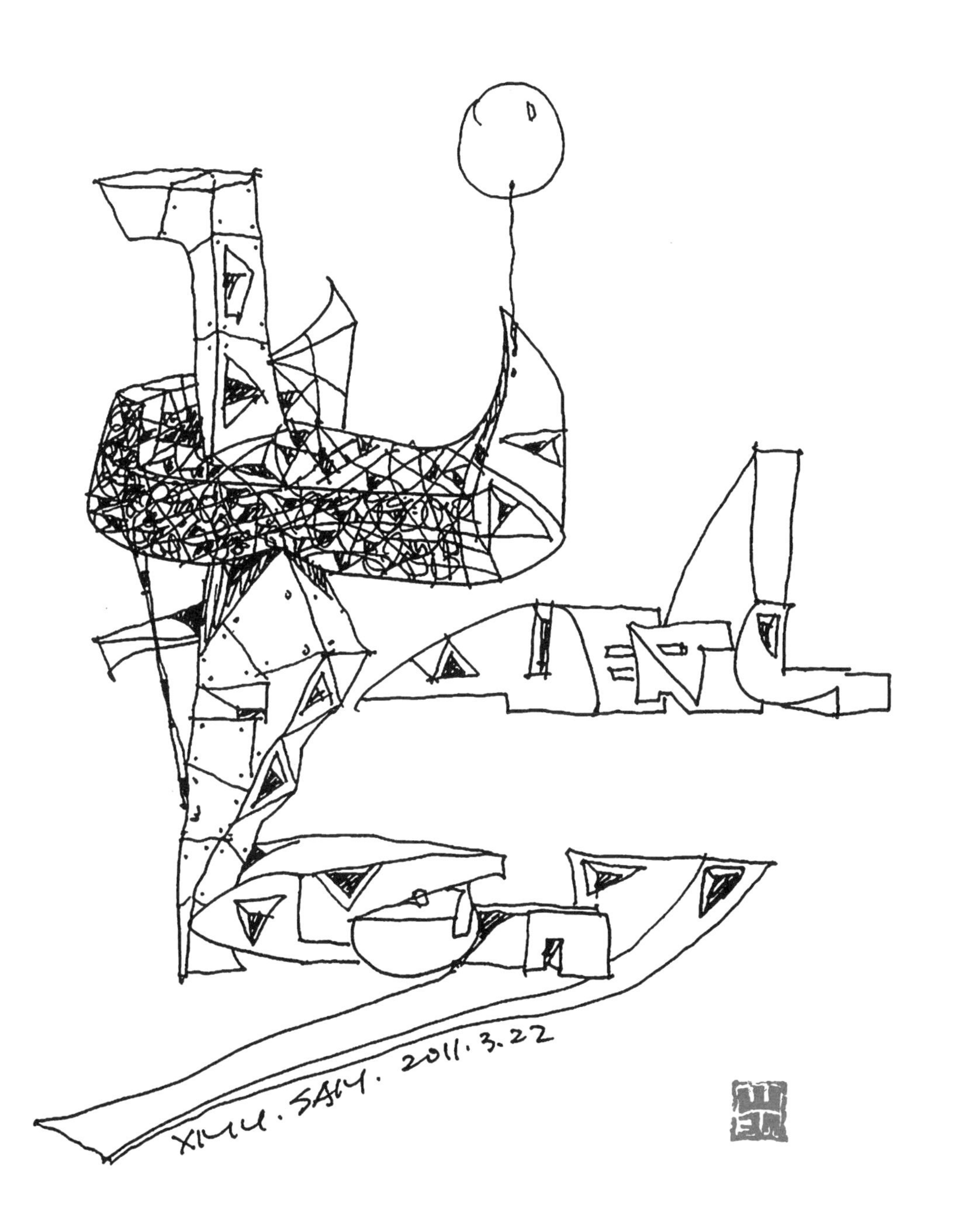
XMU. SAM. 2011.3.22

2011.3.22

XMH. 2011.3.22

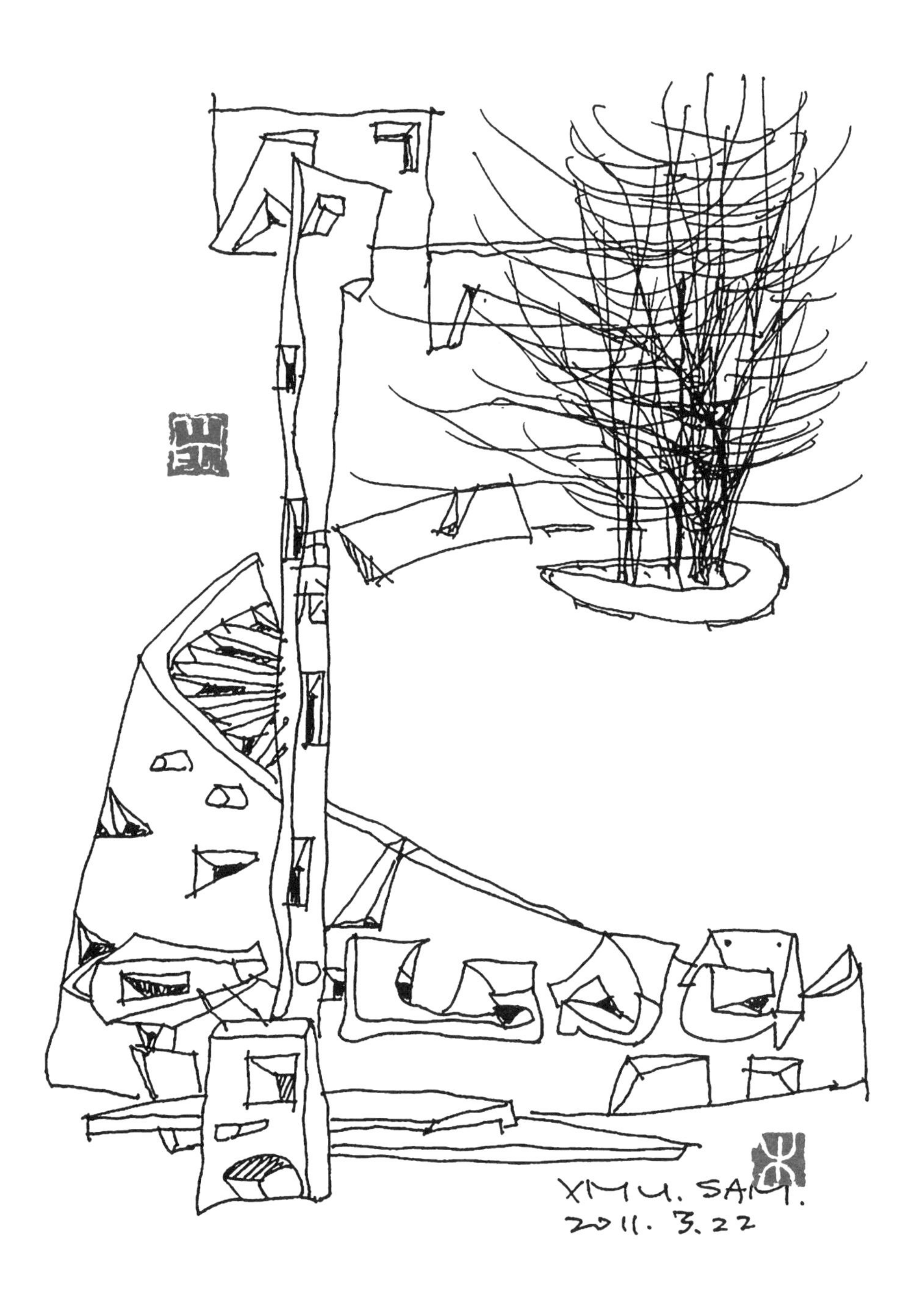
XMU. SAM.
2011. 3.22

2011.3.22
XMU.三市.

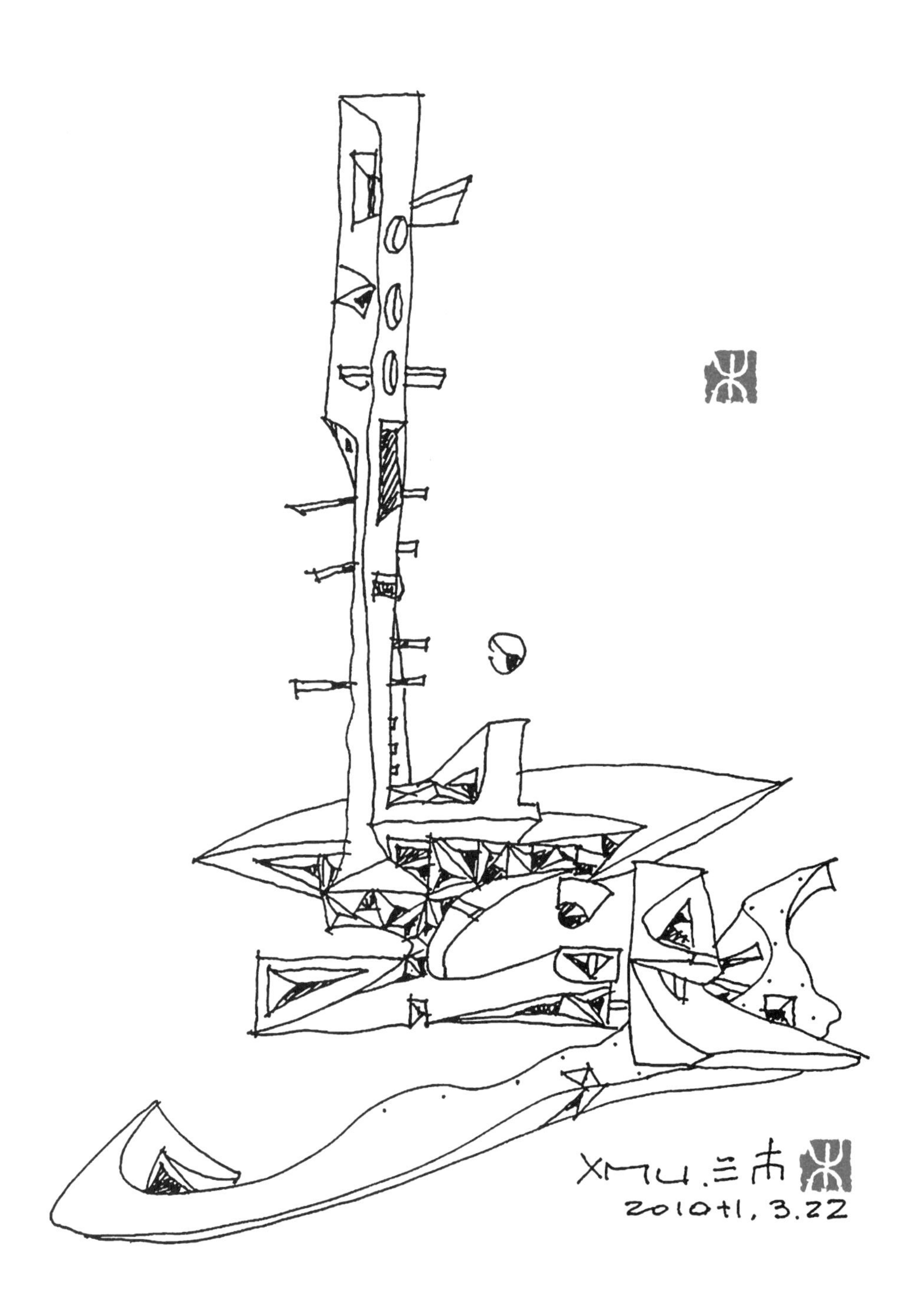
XMU.三木
2010+1, 3.22

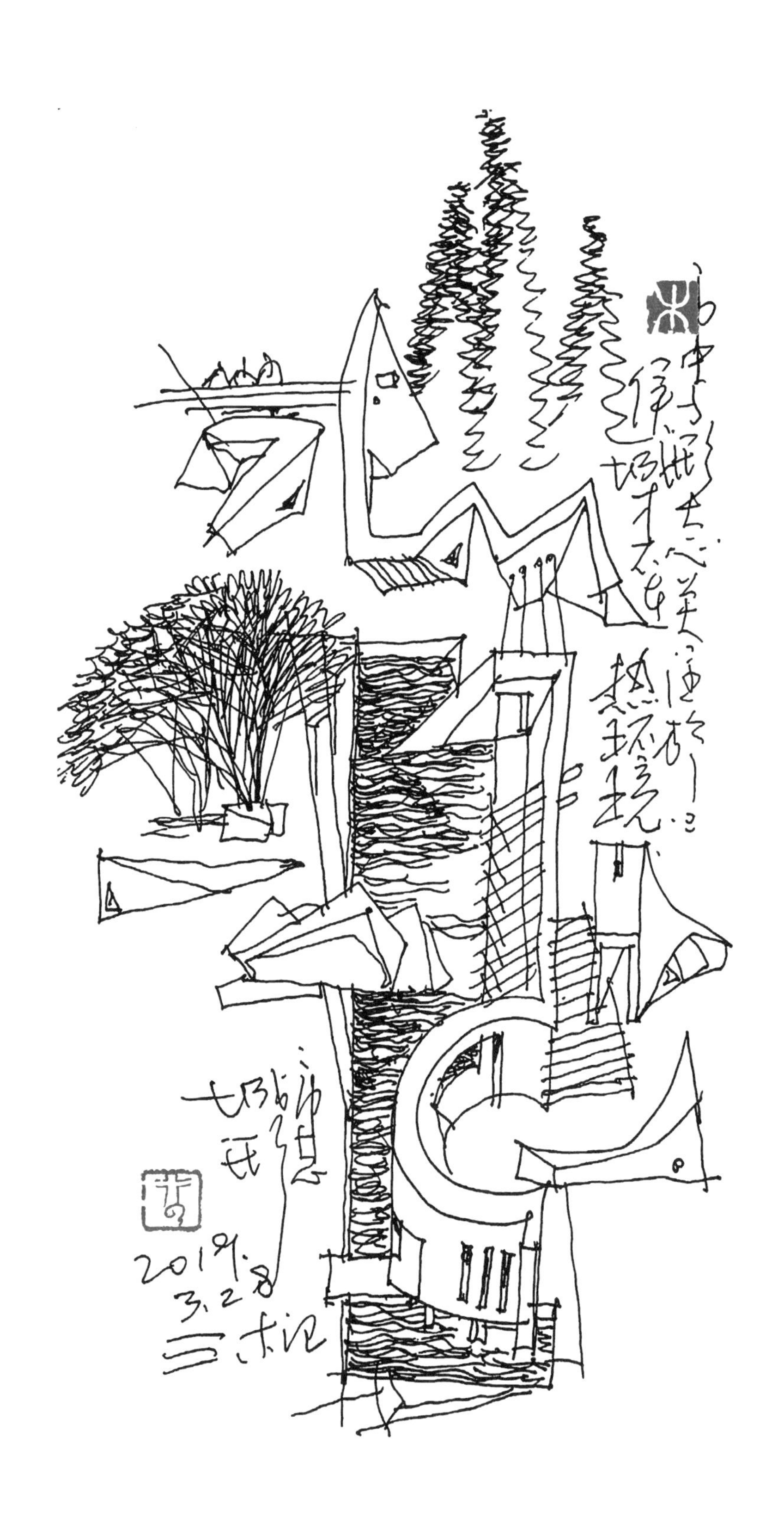
2019.
3.28

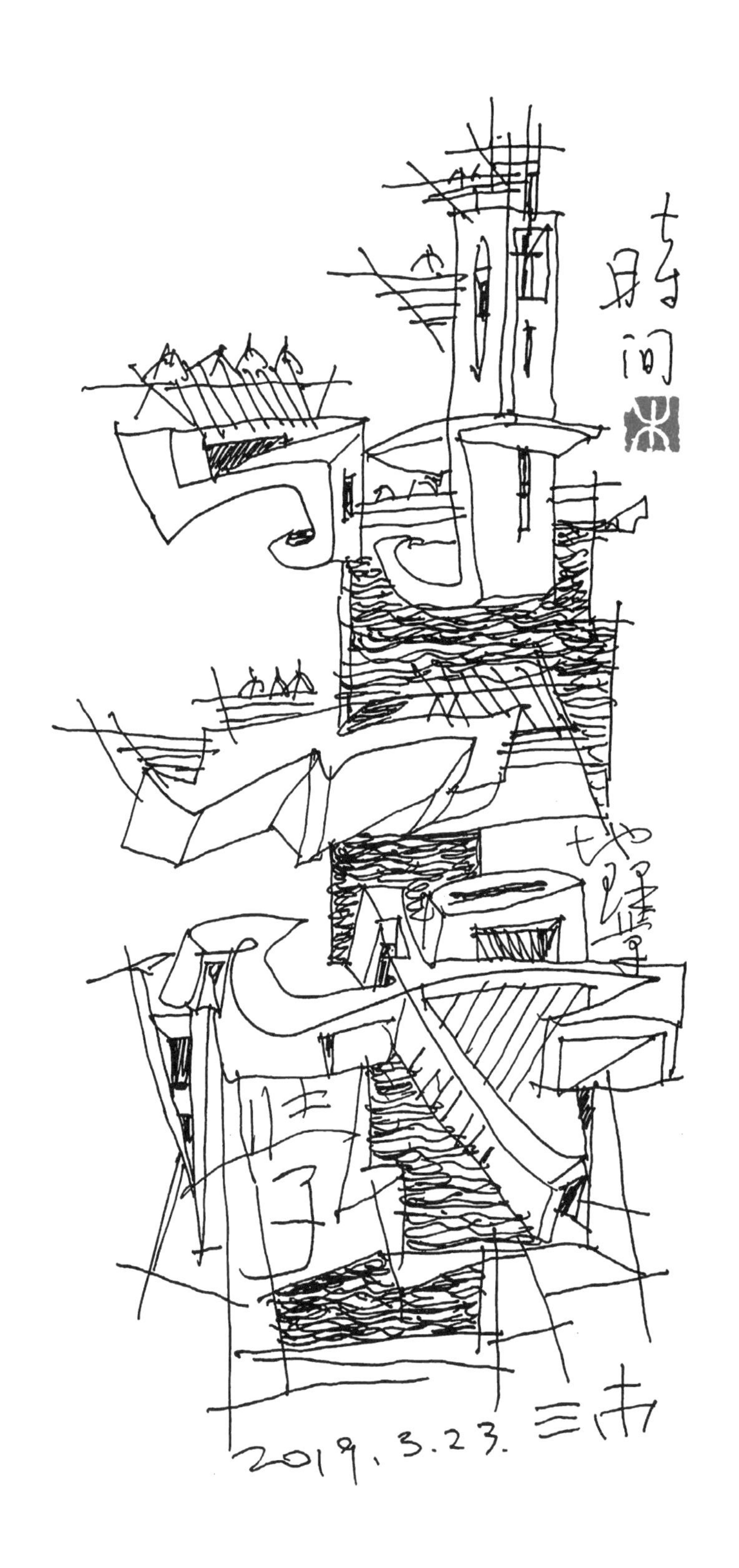
時间
2019.3.23.

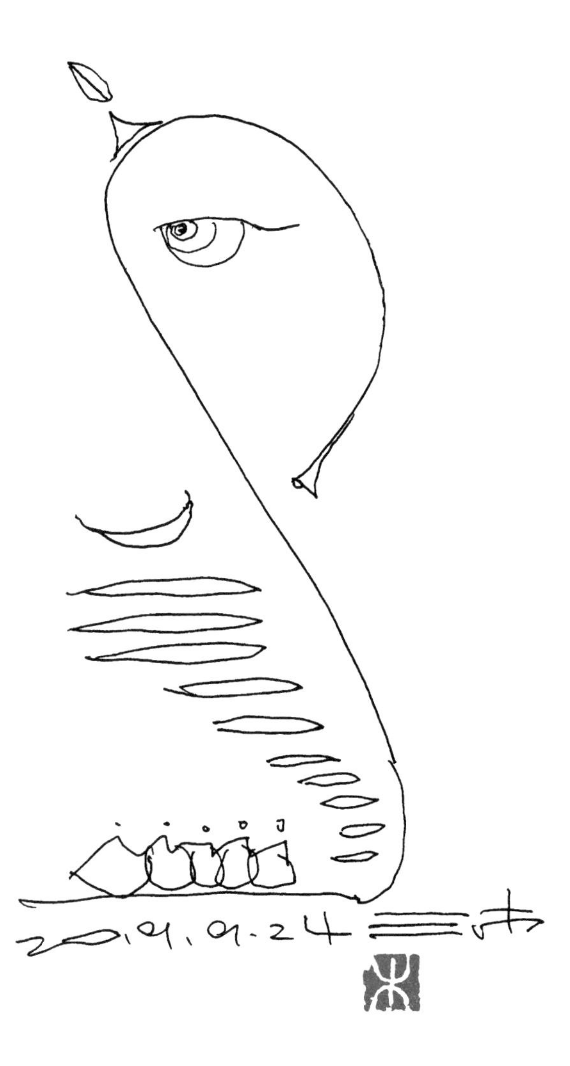
2019.09.24

3

若建筑——成果最终可提升设计能力和激发设计启迪

Pan-Architecture could inspire and increase the design.

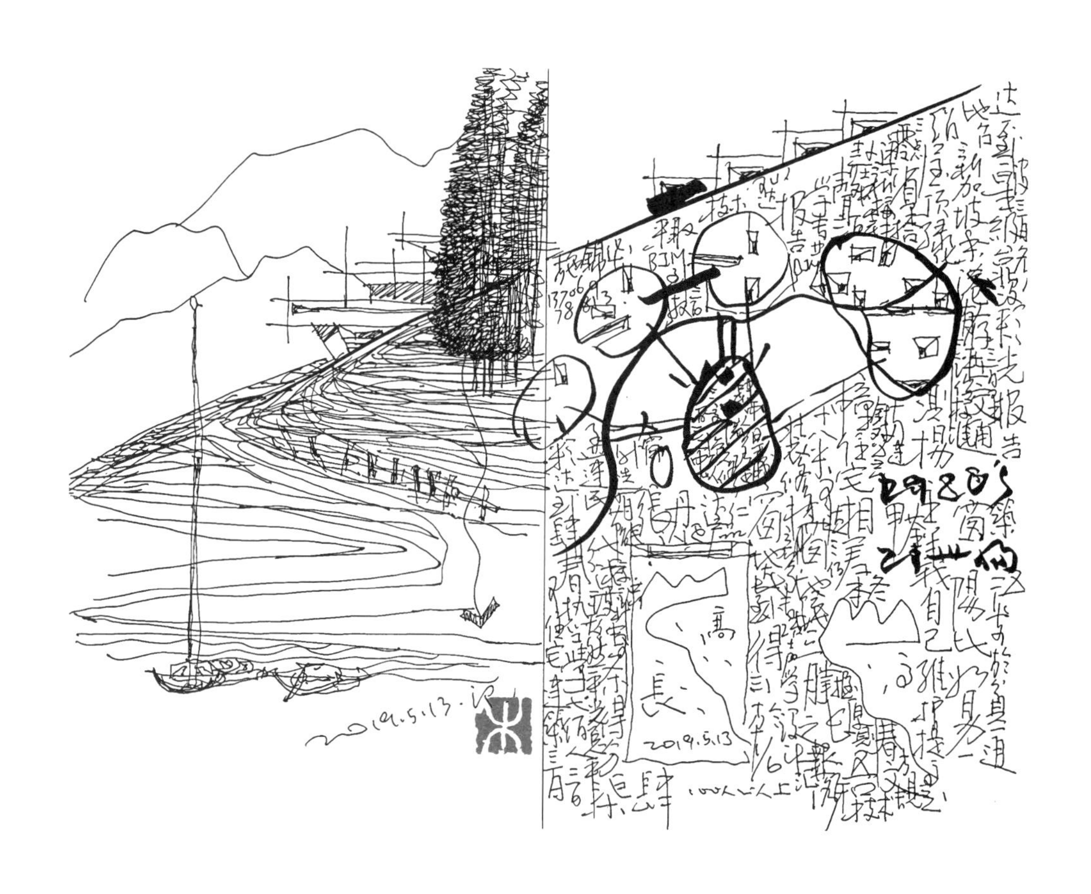

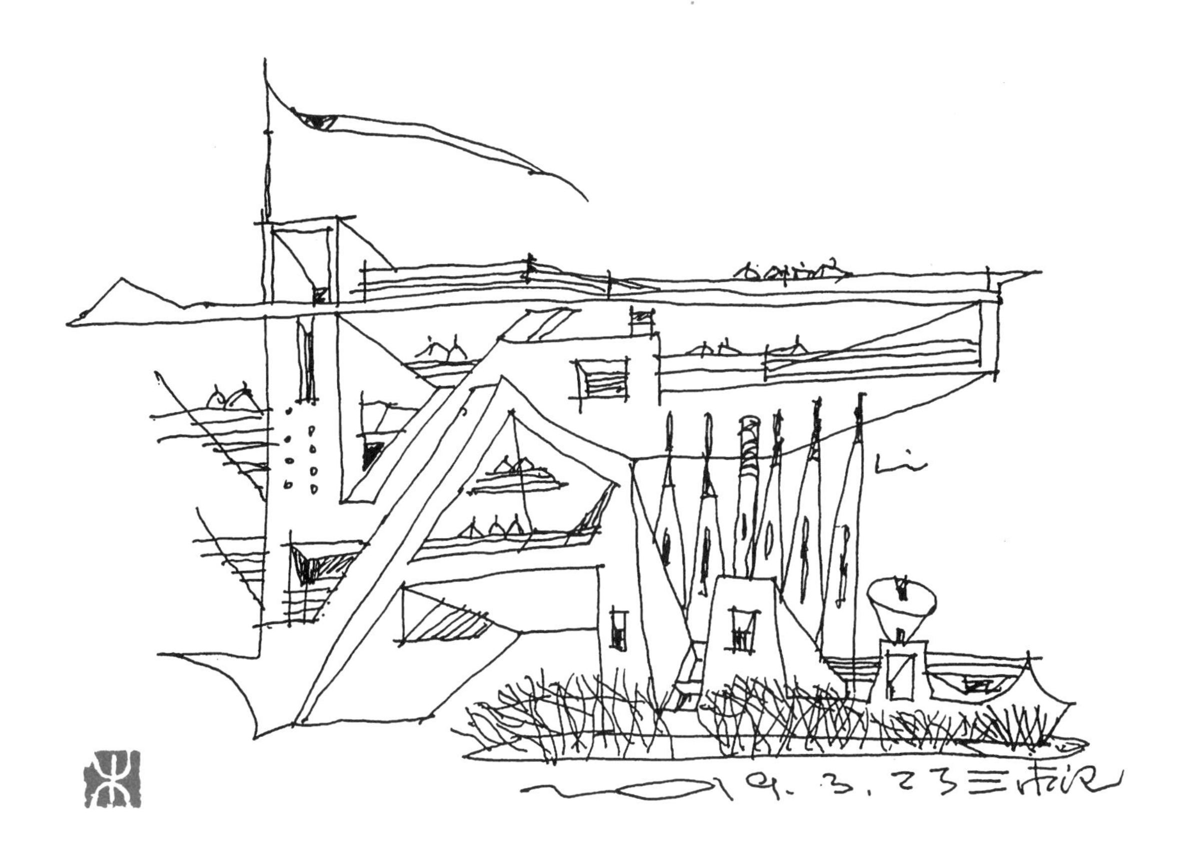

2019.3.25
2019.1.15.

2018.12.8

海港
2018.12.8

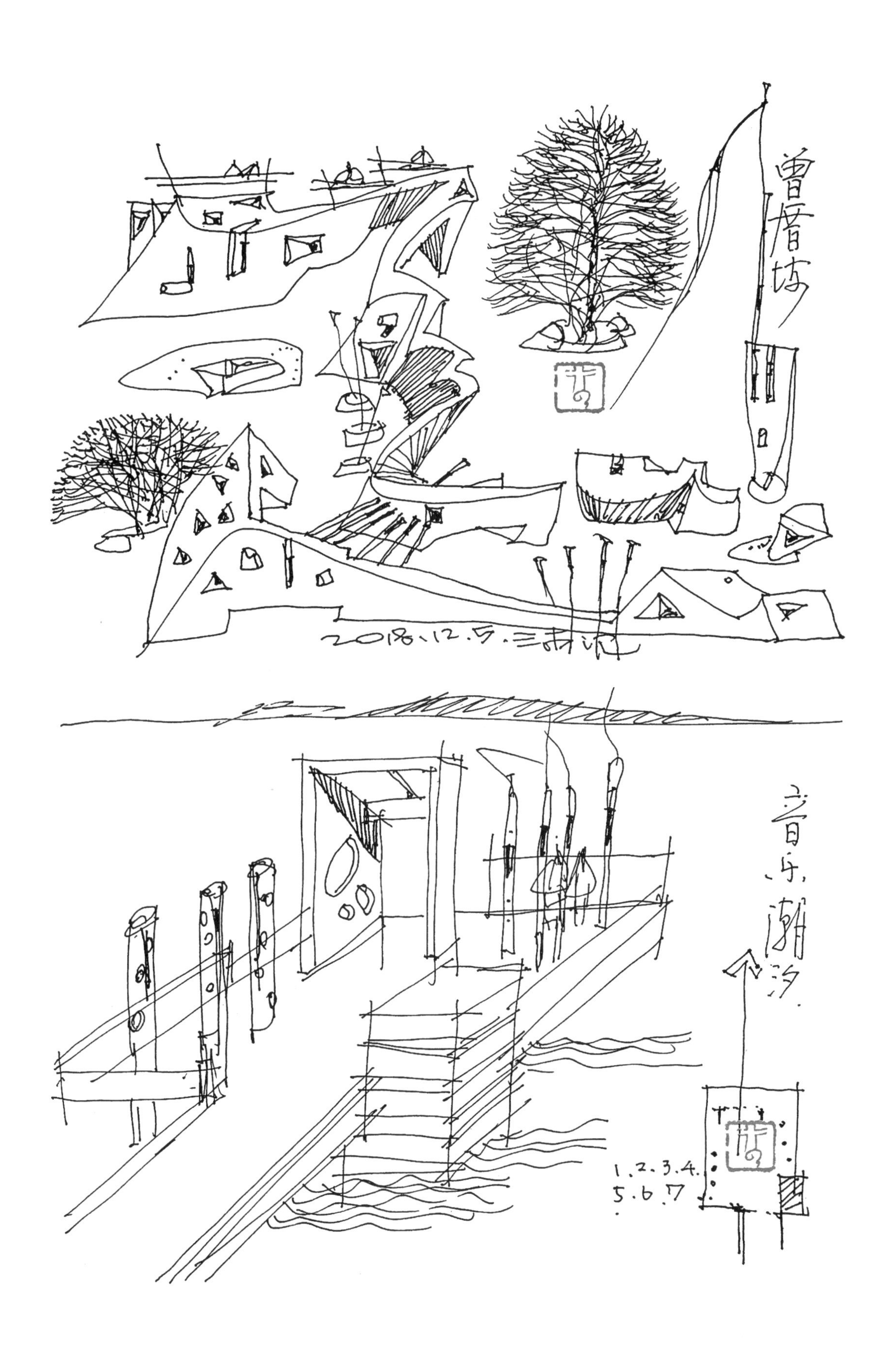
曾厝垵
音乐潮汐
1.2.3.4.
5.6.7.

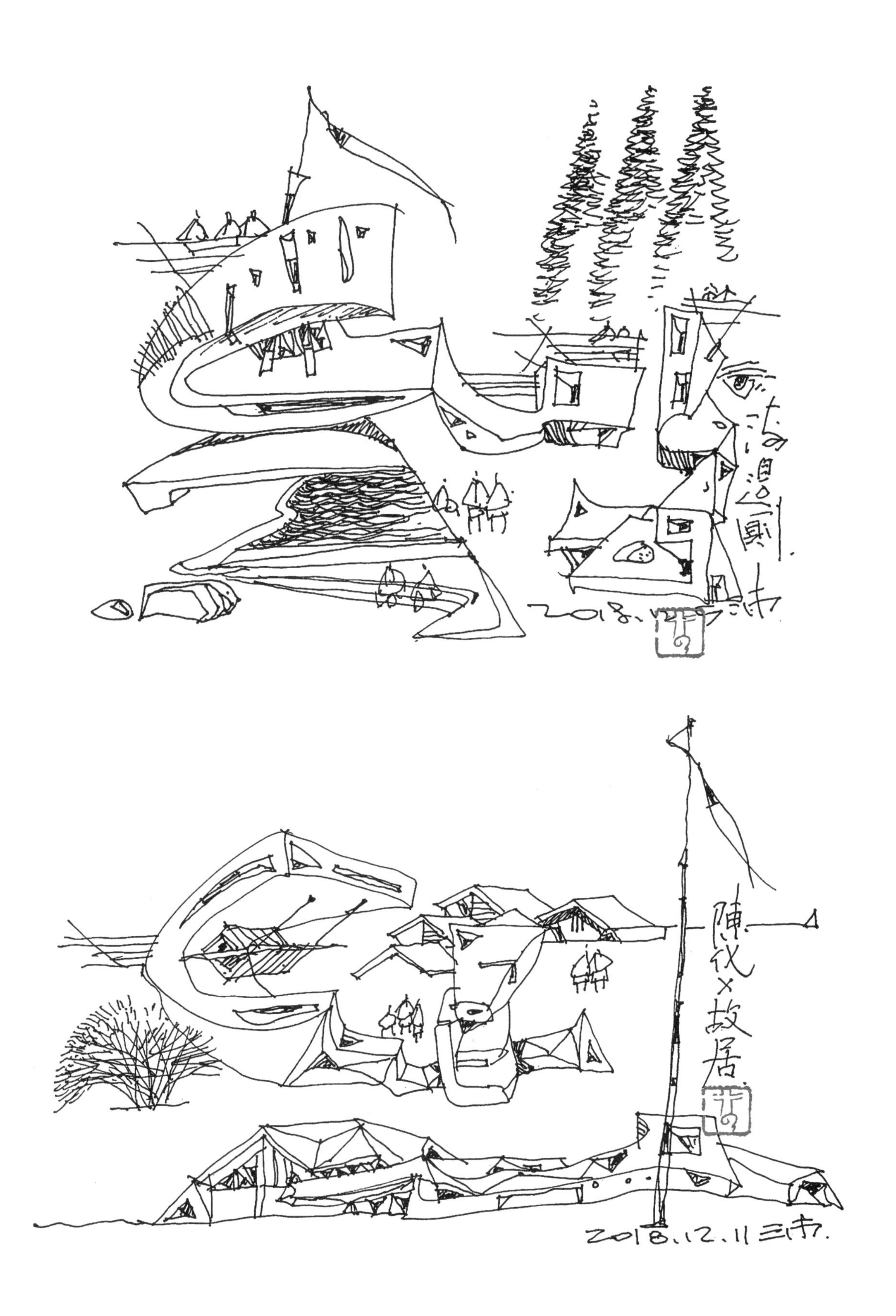
2018.12.11

空港
airport. 2018.12.8
2018.12.6

三市.
2018.12.5
坡尾
2018.12.4 三市.

2018.11.27.

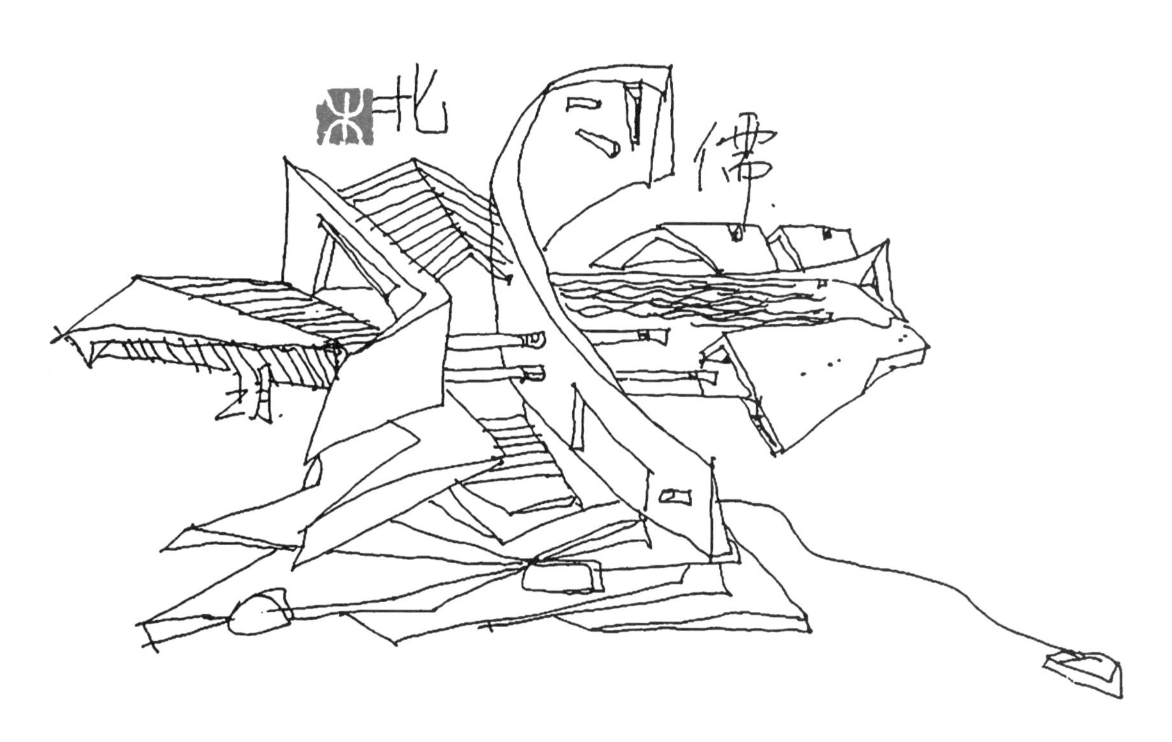
佛

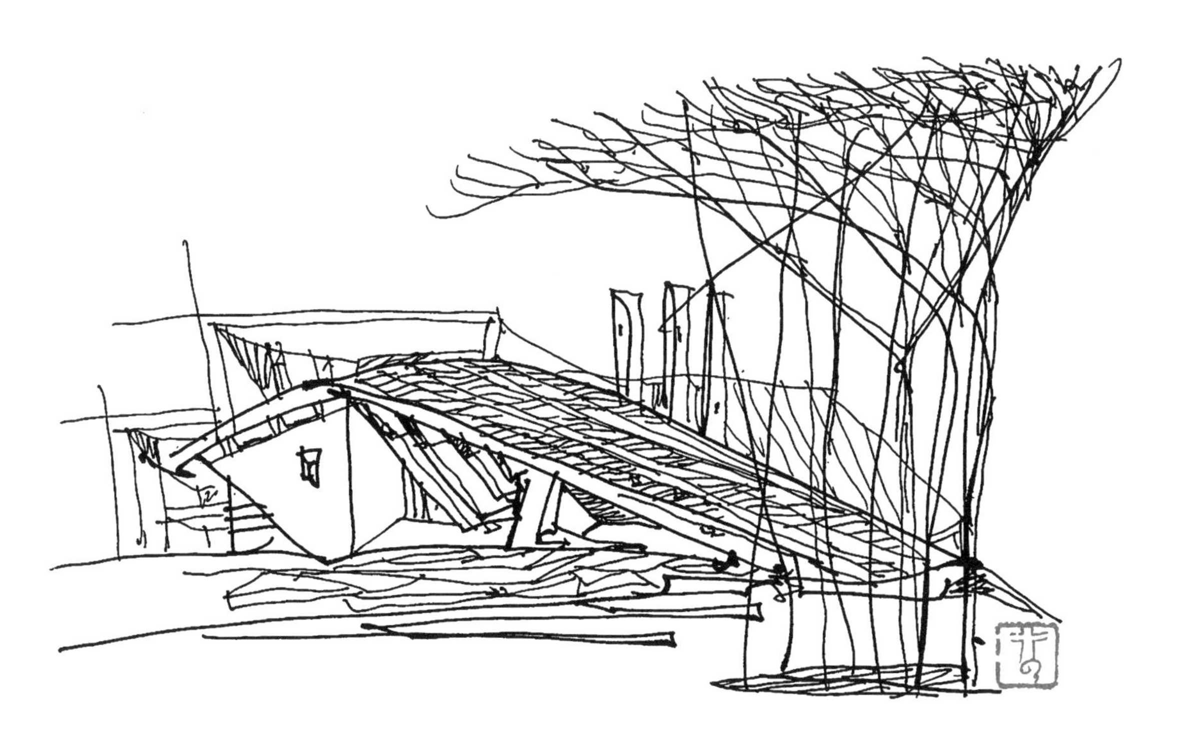

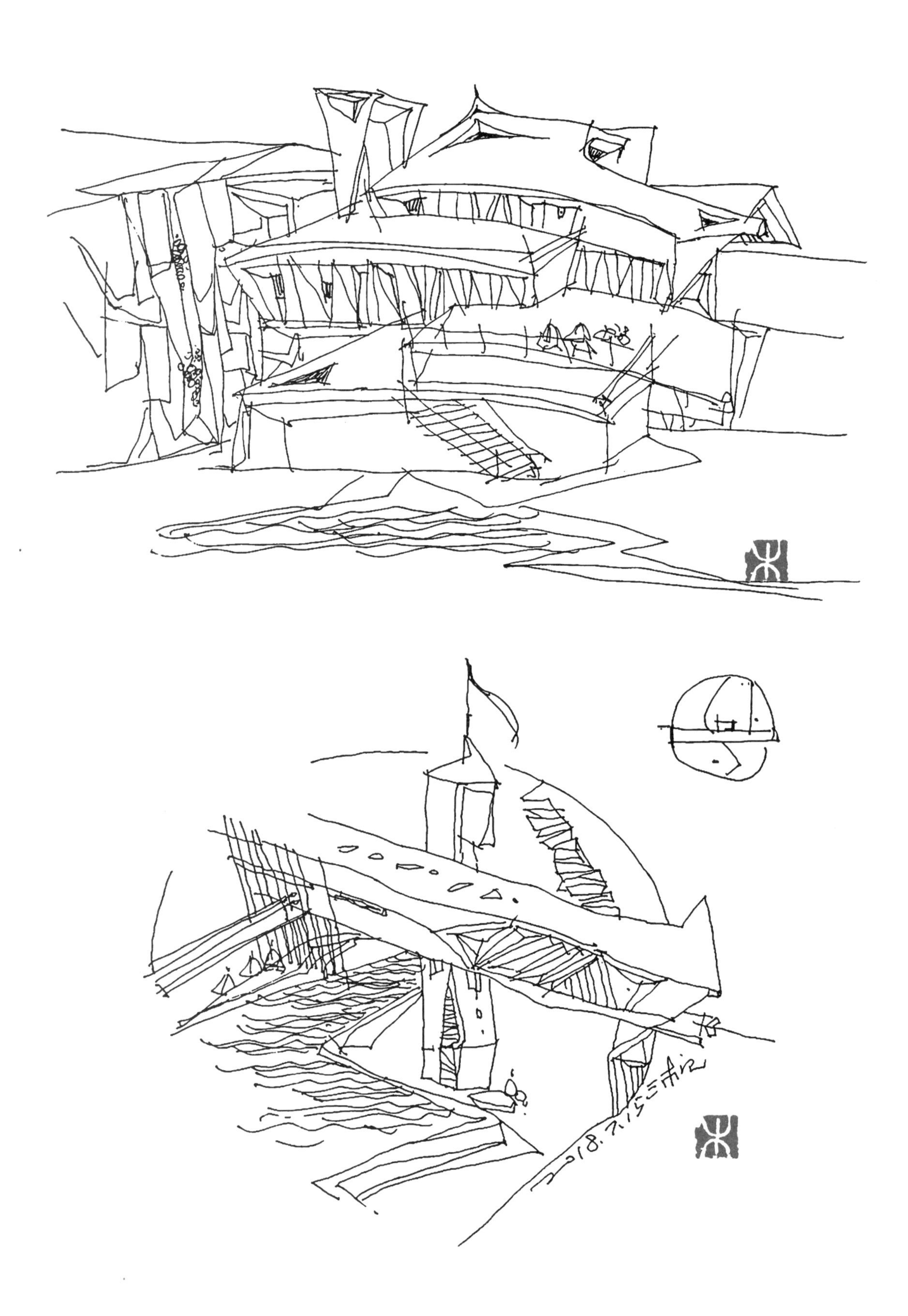

致 谢

《若建筑》至今，经过多年的点点滴滴，已形成一个建筑师的职业习惯，现在集结出版。在此特别感谢朋友们的鼓励和同事、学生的帮助！但愿它会给你带来一种不一样的图示表达和感受。

Up to now, after years of dribs and drabs, “Pan–Architecture” has become a professional habits of an architect and is now assembled for publication. I would like to express my special thanks to my friends for their encouragement and to my colleagues as well as my students for their help. Hopefully, it will give the readers a different pictorial expression and sensation.

王绍森 Wang Shaosen

教育背景：

2004—2010 华南理工大学建筑学院博士（导师：何镜堂院士）

1990—1993 合肥工业大学建筑系硕士（导师：林言官教授）

1983—1987 合肥工业大学建筑系学士

工作经历：

1987.07—1995.11 合肥工业大学建筑系教师

1995.12—至今 厦门大学建筑系教师

2013.11—至今 厦门大学建筑与土木工程学院院长

建筑学教授，国家一级注册建筑师

首届福建省建筑设计大师（2011）

当代百名中国建筑师

中国建筑教育奖获得者

社会兼职：

厦门人大第十二届常委，厦门市城市规划委员会委员，厦门市风貌建筑保护委员会委员，中国建筑学会资深会员，福建省美术家协会会员，福建省城市科学研究会副会长

Education Background：

M.A and B.A, Hefei University of Technology（HFUT, Anhui）

Ph.D, South China University of Technology（SCUT, Guangzhou）

Professional Experience：

1987.7 —1995.11, Teacher of Hefei University of Technology（HFUT, Anhui）

1995.12—pres, Processor in Architecture and Civil Engineering

2013.11—pres, Dean of Architecture and Civil Engineering in Xiamen University（XMU, Fujian）

Professor of Architecture, Class 1 Registered Architect（PRC）

Architectural Design Master of Fujian Province （2011）

Member of Contemporary 100 Chinese architects

Winner of China Architecture Education Award

Social Part-time Job：

Member of the 12th Standing Committee of Xiamen People's Congress

Member of Xiamen Urban Planning Committee

Member of Xiamen Historical Buildings Protection Committee

Senior member of Architectural Society of China

Member of Fujian Artists Association

Vice President of Fujian Urban Science Research Association